Have you done it, sir? Last time you were around, you promised you would table a motion to set up a house committee to examine the drought conditions under the chairmanship of an eminent economist.

With grateful acknowledgement to cartoonist R.K. Laxman, who has the gift of expressing profound thoughts in simple lines.

No, I won't advise you to promise them water. Promise something simple.

With grateful acknowledgement to cartoonist R.K. Laxman

BHAKRA-NANGAL PROJECT

Books in the series

Appraising Sustainable Development
Water Management and Environmental Challenges
Asit K. Biswas and Cecilia Tortajada (eds) (2005)

Integrated Water Resources Management in South and South-East Asia
Asit K. Biswas, Olli Varis, and Cecilia Tortajada (eds) (2005)

Water as a Focus for Regional Development
Asit K. Biswas, Olcay Ünver, and Cecilia Tortajada (eds) (2004)

Water Policies and Institutions in Latin America
Cecilia Tortajada, Benedito P. F. Braga, Asit K. Biswas and
Luis E. García (2003)

Integrated River Basin Management
The Latin American Experience
Asit K. Biswas and Cecilia Tortajada (2001)

Conducting Environmental Impact Assessment for Developing Countries
Asit K. Biswas and Modak Prasad (2001)

Sustainable Development of the Ganges-Brahmaputra-Meghna Basins
Asit K. Biswas and Uitto I. Juha (2001)

Women and Water Management
The Latin American Experience
Cecilia Tortajada (ed.) (2000)

Core and Periphery
A Comprehensive Approach to Middle Eastern Water
Asit K. Biswas, John Kolars, Masahiro Murakami, John Waterbury,
and Aaron Wolf (1997)

National Water Master Plans for Developing Countries
Asit K. Biswas, César Herrera-Toledo, Héctor Garduño-Velasco,
and Cecilia Tortajada-Quiroz (eds) (1997)

BHAKRA-NANGAL PROJECT
Socio-economic and Environmental Impacts

R. RANGACHARI

प्रणीति अनुसंधान केन्द्र
CENTRE FOR POLICY RESEARCH

日本財団
The Nippon Foundation

OXFORD
UNIVERSITY PRESS

OXFORD
UNIVERSITY PRESS

YMCA Library Building, Jai Singh Road, New Delhi 110 001

Oxford University Press is a department of the University of Oxford. It furthers
the University's objective of excellence in research, scholarship, and education
by publishing worldwide in

Oxford New York
Auckland Cape Town Dar es Salaam Hong Kong Karachi Kuala Lumpur
Madrid Melbourne Mexico City Nairobi New Delhi Shanghai Taipei Toronto

With offices in

Argentina Austria Brazil Chile Czech Republic France Greece Guatemala
Hungary Italy Japan Poland Portugal Singapore South Korea Switzerland
Thailand Turkey Ukraine Vietnam

Oxford is a registered trademark of Oxford University Press
in the UK and in certain other countries

Published in India
By Oxford University Press, New Delhi

ISBN-13: 978-0-19-567534-4
ISBN-10: 0-19-567534-7

Typeset in Times New Roman 10/12
by Excellent Laser Typesetters, Pitampura, Delhi 110 034
Printed at Taniya Litho-graphics India (P) Ltd., Delhi 110 092
Published by Manzar Khan, Oxford University Press
YMCA Library Building, Jai Singh Road, New Delhi 110 001

CONTENTS

Foreword by Asit K. Biswas   vii
Preface   xvii
Acknowledgements   xix
Tables, Figures, and Boxes   xxv
Glossary   xxvii
Acronyms and Abbreviations   xxxiii

1. Introduction   1

2. Indus Basin Development till 1947   11

3. Indus Basin Development after 1947   25

4. The Pre-Project Scene   38

5. Bhakra-Nangal Project   56

6. Irrigation from the Bhakra-Nangal Project   80

7. Irrigated Agriculture and Increased Productivity   95

8. Production of Food Grains   122

9. Hydroelectric Power   135

10. Flood Management   149

11. Incidental and Indirect Benefits   165

12. Sedimentation in the Bhakra Reservoir   183

13. Impact of Bhakra-Nangal Project: People's Perception   192

14. Some Important Lessons Learnt   206

    Annexures   218
    Bibliography   244
    Index   249

FOREWORD

Large dams have become very controversial in recent years. Proponents of such structures claim that they deliver numerous benefits, among which are increased water availability for domestic and industrial purposes, increased agricultural production because of reliable availability of water for irrigation, protection from floods and droughts, generation of hydroelectric power, navigation, and overall regional development which improves the quality of life of the people, including women. They argue that like any other large infrastructure development project or national policy, dams have both benefits and costs. However, the net benefits of dams far outweigh their total costs, and thus the society as a whole is much better served by their construction.

In contrast, the opponents argue that dams bring catastrophic losses to the society and the resulting societal and environmental costs far outweigh the benefits they may bring. They claim that dams further accentuate the inequalities in terms of income distribution in the society, since benefits accrue primarily to the rich, and the poor people slide further down the economic ladder. They further claim that the main beneficiaries of large dams are construction companies, consulting engineers, and corrupt politicians and government officials, who work in tandem to promote them. Poor seldom benefit from the dam 'gravy train': they mostly suffer.

The overall views of the proponents and the opponents of the large dams have been polar opposite. Scientifically and logically, these two views cannot be correct. Most unfortunately, however, there has never been a real and sustained dialogue between the two camps. For example, during the Second World Water Forum in the Hague in 2000, the pro-dam sessions mostly discussed their benefits, while the anti-dam sessions blamed all the societal ills on them, irrespective of how they were caused. The proponents and opponents of dams did not attend each other's sessions, and thus there was no debate, dialogue, and interchange of views between the two groups. Both the camps went home thinking that the Forum basically agreed with their views!

The situation was a little better during the Third World Water Forum in Kyoto, primarily because the International Hydropower Association (IHA, a pro-dam professional association) and the International Rivers Network (IRN, exclusively an anti-dam NGO) arranged a debate on the benefits and costs of large dams. By all accounts, the pro-dam group won this particular debate hands down, primarily because one of the two debaters IHA invited, Dr Cecilia Tortajada, focused her presentation on the benefits and costs of the Atatürk dam, based on observed facts and scientific analyses, and not on polemics or hypotheses. In contrast, the IRN's generalized innuendos were extensively attacked by the audience for being highly 'economical' with the truth. However, this type of interaction between the two opposing camps, though much needed, have been very rare. Such interactions must be encouraged in the future, since a societal consensus on this complex issue can be reached only through similar extensive and intensive discussions.

An important and comparatively simple question that needs to be asked and answered is why in the twenty-first century, with major advances in science and technology, is it difficult to ascertain the real costs and benefits of large dams so that their net impacts and nature of beneficiaries can be determined authoritatively and comprehensively? The current sterile and often dogmatic debate on the overall benefits and costs of large dams needs to be resolved conclusively once and for all so that appropriate water development policies can be formulated and implemented, especially in developing countries so that the overall societal benefits can be maximized. Prima facie, it should not be a difficult question to answer.

However, the world of development is complex: with scientific uncertainties, regional variations, differences in physical conditions, technical and management capacities, economic and environmental conditions, vested interests, dogmatic views, hidden agendas, and personal benefits. The issue of dams is no exception to this rule and, not surprisingly, it has fallen victim to these complex interactions of forces.

There is no question that there are many people who have a vested interest in this debate, irrespective of which side they are in. Much has been written and said about the construction and consulting companies that are associated with the planning, design, and construction of dams and their financial contributions to political parties, who are often the final arbiters of making decisions in democratic societies. There is no question that construction and development of large dams is a capital-intensive activity, and many people and groups benefit economically from this process. The dam lobby is often portrayed by the anti-dam lobby as being the 'fat cats', who are interested in the construction of dams primarily because of the financial benefits they obtain from their planning and construction processes.

An unfortunate development of recent years is the fact that some of the main stakeholders of water development projects, like farmers, have not been a part of the global debate on dams. Even during the deliberations of the World Commission on Dams (WCD), when consultations with the various stakeholders were supposed to have been carried out, one of the main stakeholders, the farmers, was conspicuously ignored during the entire process. Perhaps a balanced discussion will take place if all the major stakeholders participated in this debate.

In contrast, those NGOs who are against dams (there are numerous pro-dam NGOs as well, but they are generally not as media savvy as the anti-dam NGOs, and thus not as visible) consistently like to portray themselves as little 'Davids' who are pitted against well-heeled 'Goliaths' of the pro-dam lobby who, they claim, are highly connected to the corridors of power because of their extensive and regular financial contributions.

There is no question that there are many grass roots NGOs who have made a real contribution in highlighting the plight of the people who should have been properly resettled due to the construction of large development projects (dams, new towns, airports, highways, etc.). However, many of the main activist NGOs in the anti-dam lobby have now become financially powerful, often from the support received from several international institutions, primarily from the Western countries. Their self-portrayal as little 'Davids' is more for media and publicity purposes. Some of them have already become 'Goliaths'.

The current clout of NGOs can be realized from recent research by the Johns Hopkins Centre for Civil Society, which shows that by 1999, the non-profit sector (not including religious organizations) had become a $1.1 trillion industry in the world, employing some 19 million fully paid employees. This represented the world's eighth largest economy. In addition, a global assessment of NGOs carried out by Sustainability, a reputed NGO, in 2003 has noted that 'the NGOs that once largely opposed—and operated outside—the system is becoming integral to the system.'

The international activist anti-dam NGOs are at present a part of this overall global scene. They have become adept at playing the system to promote their own agendas and interests, at least in terms of getting funds from the funding institutions of developed countries, and generating extensive media publicity for their causes, some real, but others not. In fact, if media publicity is to be considered, the current situation is exactly the reverse: anti-dam activists have become the 'Goliaths' compared to the 'Davids' of the pro-dam lobby.

Sebastian Mallaby, an editorial writer of the *Washington Post*, reviewed two dam projects in 2004 that have been consistently opposed by the activists: a dam on the Nile at Bujagali, Uganda, and the other, the Qinghai Project in Tibet. His findings are worth noting:

> This story is a tragedy for Uganda. Clinics and factories are being deprived of electricity by Californians whose idea of an electricity crisis is a handful of summer blackouts. But it is also a tragedy for the fight against poverty worldwide, because projects in dozens of countries are similarly held up for fear of activist resistance. Time after time, feisty Internet-enabled groups make scary claims about the iniquities of development projects. Time after time, Western publics raised on stories of World Bank white elephants believe them. Lawmakers in European parliaments and the U.S. Congress accept NGO arguments at face value, and the government officials who sit on the World Bank's board respond by blocking funding for deserving projects.
>
> The consequences can be preposterously ironic. NGOs claim to campaign on behalf of poor people, yet many of their campaigns harm the poor. They claim to protect the environment, but by forcing the World Bank to pull out of sensitive projects, they cause these schemes to go ahead without the environmental safeguards that the Bank would have imposed on them. Likewise, NGOs support to hold the World Bank accountable, yet the Bank is answerable to the governments who are its shareholders; it is the NGOs' accountability that is murky. Furthermore, the offensives mounted by activist groups sometimes have no basis in fact whatsoever.

In the final analysis, there are extremists in both the anti- and pro-dam lobbies, who have their own vested interests, hidden agendas, and dogmatic beliefs. Their views and statements need to be carefully analysed in terms of their accuracy and generalizations based on limited or no facts and sometimes deliberate misstatements. Most regrettably, truth, objectivity, and concerns for the poor have often become a casualty in this partisan fight.

The sweeping generalizations of the two groups mostly do not survive any serious scientific scrutiny. In the cacophony of arguments and counter-arguments, what is invariably forgotten is that issues involved are complex and that no single conclusion can be equal for all the dams of the world, constructed or proposed, irrespective of their locations and impacts. Nor can one answer be everlasting in any specific country: it often varies over time because of changing circumstances.

What has been forgotten in the current debate on dams is that neither the statement 'all dams are good', nor 'all dams are bad and thus no new ones should be constructed' is correct. Irrespective of the criteria of 'goodness' selected, there are both good and bad dams. Furthermore, their needs vary from one country to another and often from one region to another within the same country, especially for large countries like Brazil, China, India, Mexico or Turkey, depending upon their climatic, economic, social, and environmental conditions. Countries are at different stages of economic development, and thus their need for dams could vary, depending upon their stage of development. An industrialized country like the United States has mostly developed all its best and economic dam sites. In contrast, much of the potential in sub-Saharan Africa (with the exception of South Africa) has yet to be tapped. Equally, a country like Nepal, which has a similar level of hydropower potential as the United States, has developed only about 4 per cent of its hydro potential. Thus, what may appear to be a logical and efficient policy on dams for the United States at present is unlikely to be the best and the most suitable for Nepal. In the area of dams, like in most other complex development-related issues, there is simply no 'one size that fits all'. Both the proponents and opponents of the dam debate have continued to ignore this simple fact.

A major reason for the current non-productive debate on dams is because of the absence of objective and in-depth ex-post analyses of the physical, economic, social, and environmental impacts of large dams some 5, 10 or 20 years after their construction. At present, thousands of studies exist on environmental impact assessments (EIAs) of large dams, some of which are very good but others are not even worth the paper on which they are printed. It should be noted that all EIAs are predictions and that until the dams become operational, their impacts (types, magnitudes, and spatial and temporal distributions) are not certain, and thus they remain in the realm of hypotheses. Even the very best EIAs can perhaps forecast only about 70–80 per cent of the expected impacts accurately in terms of time, space, and magnitude. For an average EIA of a large dam, often some 40–70 per cent of its impacts (positive or negative) are not properly identified.

In addition, all EIAs should include both positive and negative impacts, and not follow the present widely practised assessment of only the negative impacts. In contrast, there is at least one country which consistently includes only positive impacts to justify any project, and many of these so-called impacts are often figments of imagination of incompetent analysts. Like any major policy, programme or project, large dams have both positive and negative impacts. What is needed is a two-pronged approach which should include identification and assessment of the positive benefits and recommendations of measures which would maximize them, and assessment of the negative impacts and policy actions that should be taken to minimize them. Only such a comprehensive and holistic approach will ensure that the net benefits that will accrue to society because of any policy, programme or project can be maximized. Exclusive consideration of negative impacts, as is widely practised at present, is a fundamentally flawed procedure which will very seldom, if ever, contribute to the maximization of overall benefits to any society. Such practices only provide partial information, which will not constitute a good, logical, and scientific basis for decision making. Yet, this is the custom that is widely practised and almost universally accepted at present, not only in the field of water but also for other areas of development.

While thousands of studies are available on EIAs of large dams which were prepared prior to their construction, assessment of actual impacts of large dams 5, 10 or 20 years after their construction, from anywhere in the world, can be counted in the fingers of one's hands and still have some fingers left over! Some people have now claimed that WCD prepared numerous assessments of large dams from different parts of the world. Regrettably, most of these studies are superficial and often skewed to prove the dogmatic and one-sided views of the authors who were specially selected by WCD to prepare these studies. These studies are not objective, comprehensive or authoritative. It is possible that among these assessments, there are a few good case studies. Most unfortunately, however, no rigorous peer reviews of these case studies were ever undertaken by WCD. Consequently, the 'wheat' remain indeed very well hidden among the 'chaff'.

At the Third World Centre for Water Management, we have given considerable thought to the actual impacts of large dams from all over the world, both beneficial and adverse. We have reviewed available documents which consider impacts of such infrastructures, and we have discussed the associated complex issues with the world's leading experts from engineering, economic, social, environmental, and legal professions. We have also listened carefully to both the pro- and anti-dam lobbies.

Based on these extensive analyses and discussions, as well as visits to numerous large dams all over the world, we have concluded that the current 'dogfight' between the anti- and the pro-dam lobbies is counterproductive. However, before giving our views on this issue, a caveat is in order. Our Centre does not make a living in either promoting or opposing large dams. The activities of our Centre, which can remotely be considered to be dam-related, have always been less than 10 per cent of its annual budget. Furthermore, the Centre is totally a knowledge-based organization, and its only interest is on generation, synthesis, dissemination, and application of knowledge. The Centre eschews the concept of 'one size fits all' development philosophy for a very heterogeneous world. Based on studies carried out, the Centre has sometimes supported a dam, and equally at other times, strongly opposed construction of specific projects which were considered unnecessary.

At the Centre, we have concluded that the current controversy on dams is a dogmatic and emotional debate. If such a debate brings new issues that need to be carefully considered and addressed, it should be welcomed. However, as seems to be the case, a debate between vested interests is unlikely to be productive for society on a long-term basis. It may even be futile. The debate needs to be refocused. What is needed is the consideration of the overall architecture of water development policies, programmes, and projects which will contribute to the achievement of the objectives that developing countries desperately need: poverty alleviation, employment generation, regional income redistribution, and environmental conservation. Within this overall architecture, what is imperative is to decide how best to meet the water needs of a society in a cost-effective, equitable, environment-friendly, socially acceptable, and timely manner. The world of development is complex, and thus there will always be tradeoffs because of implementations of specific policies, programmes, and projects. These tradeoffs should be considered objectively, accurately, honestly, sensitively, and in a socio-politically acceptable manner. Within such an overall framework, the best solution for water development must be sought for each specific case. This may warrant construction of a dam in a specific location, but it may require another approach.

In the final analysis, alternatives selected may require construction of properly planned and designed dams, which could be large, medium or small, or rainwater harvesting, or any number of appropriate 'soft' alternatives like water pricing and focus on efficient management practices. The solutions selected must not be dogmatic and should always reflect the needs of the people in the areas under consideration. In terms of water development, it should be remembered that small is not always beautiful and big is not always magnificent. Solutions must be very specifically designed to solve the problems encountered. The current emotional debate on the dams is somewhat akin to a solution-in-search-of-a-problem approach, where the a priori solution becomes dams or no dams, depending upon the people and institutions concerned. Such priority solutions are often scientifically unacceptable, economically inefficient, socially disruptive, and environmentally dangerous.

The Centre has concluded that no single pattern of water development is the most appropriate solution for all the countries of the world at any specific point of history. Countries are at different stages of development;

their technical, economic, and management capacities are not identical; climatic, physical, and environmental conditions are often dissimilar; institutional and legal frameworks used for water management differ; and their social and cultural conditions may vary significantly. Thus, there is simply no one single path to water development which could be followed by all the countries of the world in perpetuity.

In addition, the world is changing very fast, and thus water management concepts and practices must change as well. The world of water management is likely to change more during the next 20 years, compared to the past 2000 years. Thus, past experiences can only be of limited help in the future, especially as most of these changes, unlike in the past, will come from outside the water sector. Among these numerous driving and overarching forces that are likely to affect water management, are rapidly changing demographic dynamics of the world, simultaneous urbanization and ruralization in developing countries, accelerated globalization and international trade, advances in technology, and continuing communication and information revolution. All these and other associated changes will affect water management through myriads of interconnected pathways, some of which can be anticipated at present but others are still mostly unknown. In this rapidly changing complex landscape of water management, there should be no room for sterile debates on large dams, as has been the case in recent years.

The main question facing the developing countries of Asia, Africa, and Latin America is *not* whether large dams have an important role to play in the future, but rather how best we can plan, design, and construct them where they are needed so that their performances in economic, social, and environmental terms can be maximized, adverse impacts can be minimized, and simultaneously ensure that those who may have to pay the costs of their implementation are explicitly made their beneficiaries. It will not be an easy task to accomplish, but it is nevertheless an essential one that must be undertaken. This can only be done if the benefits and costs of the large dams are objectively, reliably, and dispassionately assessed, and the lessons learnt (both positive and negative) are used for planning and construction of new dams, and management of the existing ones.

In order that proper lessons can be learnt, the Third World Centre for Water Management decided to conduct comprehensive impact assessments (positive and negative) of three large dams that have been operational for a minimum of 10 years. The projects selected are Aswan high dam in Egypt, Atatürk dam in Turkey, and Bhakra-Nangal Project in India. The study is being supported by the Nippon Foundation of Japan.

All the three studies include the perceptions of the people of the areas that were affected by the dams, both beneficiaries as well as those who had to pay some costs. The results of these in-depth studies have started to become available from early 2005 and these are being made available for examination by any interested party. The studies are based only on facts and scientific analyses and not on dogmas, biases, hypotheses or hidden agendas.

The present study on the impacts of the Bhakra-Nangal Project was carried out by Prof. R. Rangachari of the Centre for Policy Research, New Delhi, India. I have had the pleasure of knowing and working with Prof. Rangachari for over 25 years. He is not only one of the eminent and most knowledgeable water experts of India, but is also a very rare individual who is objective, has no hidden agenda, draws conclusions only after conducting proper scientific and technical analyses based only on observed facts and figures, and never has any hesitation in calling a spade a spade. The Centre was very fortunate to have a person of Prof. Rangachari's stature, eminence, integrity, and objectivity to conduct this definitive study. As is his style, Prof. Rangachari has produced an excellent analysis within the limited budget and time frame available.

The Centre is very grateful to Prof. Charan Wadhwa, former President of the Centre for Policy Research, during whose tenure the study was carried out. Prof. Wadhwa took a personal interest in the study, and followed it from the beginning to the end.

We would also like to express our appreciation to Dr C. D. Thatte, former Secretary of the Ministry of Water Resources, Government of India, for the special interest he took in this study from its inception to its end, as well as to Mr M. Gopalakrishnan, Secretary General of International Commission on Irrigation and Drainage, and Professors George Verghese, R. Ramaswamy Iyer, and Pratap Bhanu Mehta of the Centre for Policy Research, for their interest and support in the study.

No research project can be undertaken without appropriate financial and intellectual support. The funding for the overall study was provided by the Nippon Foundation of Japan, a unique institution dedicated to improving the quality of life and the environment throughout the world. I would like to express our appreciation and gratitude to Mr Reizo Utagawa, former Managing Director of the Nippon Foundation, who was instrumental in launching the entire project and who followed it with keen personal interest. The same interest and support was extended by his successor, Mr Shuichi Ohno. We are also grateful to Mr Masanori Tamazawa for his continuing advice, interest, and support to the overall project. Mr Tamazawa also participated in the workshop that was organized in Delhi to review the first draft of this report. Without the support of all these individuals, a study of this quality simply could not have been prepared.

ASIT K. BISWAS
President
Third World Centre for Water Management
Atizapán, Mexico

PREFACE

The recent general election in India has demonstrated, beyond any doubt, that the one issue that occupies centre stage in most states is water. It is a subject that has agitated not just rural agriculturists but also people in the cities and towns. Another issue of near-equal importance to the common man and the elite in India, is that of electric power, which is also linked to water resource development in many ways. When freshwater resources are limited and unevenly distributed over time and space, the storage of water for meeting demands round the year becomes necessary. Thus, India has built dams—large and small—over centuries, but particularly after Indian Independence.

In the past decade or two, certain articulate groups have criticized the tendency to build more dams on the grounds that their social and environmental costs have been devastatingly high. It would be possible to examine the tenability of this allegation if there were enough records and published information relating to the actual performance of the hundreds of such schemes already completed and serving the national needs. Unfortunately, there is a dearth of such credible analyses that would enable a dispassionate scrutiny of the various allegations against dams.

Almost all schemes involving dams have been built in the public sector by one governmental agency or another. Governments all over the world are very selective in revealing data and information collected over the years but buried in their files. India, too, is no exception to this general pattern. Thus, it is not surprising that hardly any independent and reliable studies of completed schemes are available in India. On the other hand, attempts by non-governmental organizations (NGOs) and private individuals to obtain the relevant data in order to conduct studies on their own, have also not been too successful. This has led to the proliferation of conjectures and patently incorrect information, which, by and large, depict a bleak record of dams.

The Third World Centre for Water Management (TWCWM) should, therefore, be congratulated for initiating a research study of selected dams in different developing countries and for their identifying India as an important nation in this regard.

It is a privilege to have been asked by the TWCWM and the Centre for Policy Research (CPR) to undertake this study on the Bhakra-Nangal Project, which has completed over fifty years of useful service in irrigation, power, and related sectors. I have endeavoured to make a rigorous, independent, and comprehensive study and trust that this effort fills an important void in the existing literature on water resource development. I hope that I have justified the faith and confidence reposed in me.

I owe gratitude to a large number of institutions and individuals, without whose help I would not have been able to complete the study.

New Delhi   R. RANGACHARI
November 2005

ACKNOWLEDGEMENTS

This book would not have been written but for the encouragement and help of a multitude of people and organizations. Foremost among them is the TWCWM and its president, Dr Asit Biswas, a friend for many decades. Dr Biswas urged me in 2002 to assess the performance of a dam project completed some decades ago and highlight its impact on the people of the region. We agreed that the Bhakra project was suitable for this purpose and he pressed me to undertake this task. Both he and the vice president of the TWCWM, Dr Cecilia Tortajada, constantly provided encouragement while allowing me complete freedom to structure this report according to my own views. In addition, the Nippon Foundation facilitated the study by its general support and assistance to the TWCWM in undertaking studies worldwide.

Writing a book of this complex nature and broad scope required the help of many institutions. Moreover, permission was needed to access official documents, as well as unpublished information, usually buried in the labyrinths of the government. I acknowledge my indebtedness to the Ministry of Power and the Bhakra Beas Management Board (BBMB) for willingly assisting me in this task. Similarly, the Ministry of Water Resources, the Central Water Commission, the Planning Commission, and the government agencies in the states of Haryana, Himachal Pradesh, Punjab, and Rajasthan, gave their cooperation by making data available and according permission to visit the sites.

It would be difficult to list in full the very large number of officials who helped me. I must, however, gratefully mention the departmental seniors who made this possible. Rakesh Nath, the dynamic chairman of the BBMB and its members are foremost among them. Balbir Singh, Member (I) and S.K. Duggal, former member, gave their support and assisted me with information about the project. Anil Arora, Secretary, also offered his cooperation in every possible way. I must record my thanks to K.G. Aggarwal, former secretary. I owe a debt of gratitude to Mahajan, Member (Power),

Jagbans Singh, Financial Advisor and Chief Accounts Officer (FA&CAO), Rajeev Bansal, Executive engineer (EE), and Raman Bajaj. H.C. Chawla, Director, BBMB, at the Delhi office helped coordinate with them.

I am grateful to the officials of the Bhakra Dam Organization at the Nangal township, who shared their experience and information with me. D.S. Notra, Chief Engineer, who headed this organization, shared with me his wisdom about the conditions in the canal command area in Rajasthan, where he had served earlier. The late J.K. Bhalla and N.K. Bhatia, superintending engineers of the Bhakra Dam Circle, and S.N. Bhatnagar, Addl. Superintending Engineer (Reservoir Management), were of immense help whenever I visited the project works and reservoir area.

N.K. Jain, Director (Designs), shared details about the Bhakra dam with me and also enlightened me about the functioning of the canal system in Punjab, where he had served earlier. Tarlochan Singh, Senior Design Engineer (SDE) was a cheerful aide in digging out archival information on the project from the extensive library. I had long sessions with K.V.S. Thakur, Director (Regulation) and his senior aides, R.K. Garg and S.P. Singh, about flood management and water allocations. They also explained the intricacies of the interstate issues of this project to me. I express my thanks to them for their help and consideration.

P.P. Wahi, Chief Engineer (Power Generation) and S.M. Dhir, Superintending Engineer (SE), freely provided information on the power aspects of the project, which I acknowledge with thanks. The chairman, BBMB, himself an expert on power generation, transmission and grid operations, had many discussions on these aspects with me, which I gratefully acknowledge.

The Ministry of Water Resources was a strong supporter of my study. I must record my sincere thanks to the former secretaries, Z. Hasan, B.N. Nawalawala, and A.K. Goswami, for their many helpful suggestions. I thank V.K. Duggal, the present secretary, for his courtesy and support. Radha Singh, former additional secretary (and presently Secretary, Agriculture) and I have had many long discussions on the various impacts of large dams and these were greatly helpful in shaping my present study.

I would like to specifically acknowledge my parent organization, the Central Water Commission (CWC), for the extensive help it gave me. Most officers there had been my colleagues in the past and they continued to extend the same regard and support that I had earlier enjoyed. R. Jeyaseelan, Chairman, and the members of the CWC, S.K. Das and M.K. Sharma, shared with me whatever information the Commission had about the project. Former chairman, A.D. Mohile, and former chief engineers, M.S. Menon and S.C. Sud, were very helpful and offered many useful suggestions. I also, thank B.D. Pateria, Chief Engineer, H.R. Sharma, Superintending Engineer, and T.K. Sadhu, Director, for their help.

M.L. Baweja, former chief engineer (CWC), served as my associate in conducting random field interviews with the project-impacted people. His diligent work formed the basis for the chapter on the people's perception of the project's impact. I am grateful to him for his assistance. His field visits were facilitated by the cooperation of many officers of the state governments, mentioned subsequently.

R.P. Yaduvanshi, chief engineer (Canals), Punjab, gave me the fullest help and cooperation in visiting the canal command areas in Punjab. Among his senior officers, I must particularly mention the help given by Harbans Singh, S.E., Patiala, and K.K. Garg, Executive Engineer (Regulation).

Similar help was extended by J.S. Ahlawat, Engineer-in-Chief, Haryana, and his executive engineer (Regulation), S.S. Choudhry. Likewise, the officers in charge of the field circles and divisions at Hisar/Sirsa/Fatehabad assisted my associate, Baweja and enabled field surveys in their respective areas. In particular, our thanks are due to Aslam, Goel, R.K. Rajpal, C.L. Ganda, Uddar, and K.G. Bansal. Deepti Umashankar, Deputy Commissioner (Hisar) and her staff assisted with information on the resettlement and rehabilitation (R&R) aspects of the project. They also facilitated the interviews with the resettled people.

N.K. Jasrai, Chief Engineer, Rajasthan, and his officers helped during the field visits of Baweja to areas in Ganganagar/Hanumangarh. B.S. Rajvanshi, former chief engineer, Rajasthan and later Consultant, International Commission on Irrigation and Drainage (ICID) assisted me with many details of the conditions in the Rajasthan Command Area, before and after the project. My grateful thanks to them.

I must acknowledge and express my gratitude for the help that many non-official international and national organizations gave me. C.V.J. Varma, former president, International Commission on Large Dams (ICOLD) and presently Secretary General, Council for Power Utilities, shared with me his experience regarding large dams and their impact, worldwide. I am also very grateful for his comments on my draft report. M. Gopalakrishnan, Secretary-General, ICID, who was my colleague in the CWC, provided much support by sharing information and commenting on my draft report. His predecessor, Dr Thatte, another old associate of mine, gave me special assistance throughout the project study. We spent long hours in various discussions and he suggested many improvements. My sincere thanks are due to him.

M.S. Menon, Member-Secretary, Indian National Committee on Irrigation and Drainage (INCID), generously helped me in several ways with information and suggestions. Along with S.C. Sood, he took a leading part in helping finalize the questionnaires for ascertaining people's views. Dr Thatte, Menon, and Sood spent many hours in discussions with me on their format and contents. I am grateful to them for this help.

Dr B.P. Dhaka, Secretary General, PHD Chamber of Commerce and Industry furnished a lot of useful information with regard to the project and its impact on industry in the area, which is gratefully acknowledged.

The draft study report was discussed at a workshop held on 12 January 2004, at the Centre for Policy Research (CPR) with a peer group. I wish to acknowledge my gratitude to all the participants who reviewed, commented, and offered suggestions for additions and improvements. A list of the participants is included elsewhere in the report.

I also received help from some others who reviewed my draft and made suggestions for improvements. I would like to particularly mention, with appreciation, the help and support I received from my daughter and critic-in-residence, Devika, as well as my son, Dilip.

I must express my debt of gratitude to the CPR for providing me with an office and giving me administrative and logistical support during the course of this study. Dr Charan Wadhva, formerly president, CPR, gave me constant encouragement. Dr Pratap Bhanu Mehta, President, CPR, also supported me in every possible way. The stimulating discussions I often had with my long-time CPR colleagues, particularly B.G. Verghese and Ramaswamy R. Iyer, were of great help in sharpening my focus and manner of presentation. I acknowledge with thanks the wholehearted cooperation rendered by the CPR staff including Dr N.K. Paswan, V.K. Bhal, Kamal Jit Kumar, Y.G.S. Chauhan, L. Ravi, Jagmohan Chander, Pradeep Khanna, M.C. Bhatt, Pramod Malik, and Sonia Bhutani.

Secretarial assistance was given by Vinodini Ramachandran, who cheerfully helped by word-processing the repeated refinements of the draft report. I wish to express my appreciation and thanks for her help.

Expert help in copyediting was rendered by Suhasini Iyer and Devika. I am grateful to both for suggesting many improvements in the presentation, in addition to the rectification of errors. I am grateful to Oxford University Press for many useful editorial suggestions.

Last but decidedly not the least, I must acknowledge and thank the patience and understanding shown by my family members. My wife, Vasantha, took complete charge of the household, thus freeing me for the Bhakra study. Akshay, Dipika, and Gautam helped in their own ways. Even young Uttara was understanding enough to relinquish her claims on my time, leaving me free to study Bhakra, whenever needed.

To all others who helped me and whom I have failed to mention here, I express my gratitude.

R. Rangachari

TABLES, FIGURES, AND BOXES

Tables

2.1   Average annual flows in the rivers of the Indus system   12

4.1   Rainfall Averages in select districts of Punjab and Haryana   43

4.2   Temperature at select locations in Punjab and Haryana (1999)   44

4.3   Mean monthly evaporation in the Beas-Sutlej catchment   45

4.4   Average wind speed (1999)   45

4.5   Mean Relative Humidity (1999)   46

4.6   Climate Statistics for Ganganagar   46

4.7   Development of irrigation in post-Partition Punjab (1950–1 to 1954–5)   48

4.8   Yields of crops in Punjab 1950–1 to 1954–5   49

4.9   Plant species in the catchment   50

4.10   Test results of water samples collected from the Bhakra dam region   52

5.1   Details of land acquisition for Bhakra submergence area   63

5.2   Rehabilitation of Displaced Families   64

6.1   BNP—New Command Area   85

6.2   Irrigation Development   89

6.3   Waterlogged areas in irrigation projects in Haryana, Punjab, and Rajasthan   92

7.1   Progressive Land Use in Punjab and Haryana   103

7.2   Area, Production, and Yield of Wheat in Punjab   105

7.3   Area, Production, and Yield of Rice in Punjab   106

7.4   Area, Production, and Yield of Wheat in Haryana   107

7.5   Area, Production, and Yield of Rice in Haryana   108

7.6   Progressive production of food grains   110

7.7   Progressive improvement in yield of major food crops   112

7.8   Sex ratios in some states of India, 2001   117

8.1   Growth of Area, Production, and Yield in food grains   125

8.2   Area under food grains and Production (1950–1 to 1993–4)   126
8.3   Annual compound growth of area, production,
      and productivity   127
8.4   Projections of the National Commission   129
8.5   Projections by the NCIWRD Plan   130
8.6   Land use in Punjab and Haryana (2000)   132
8.7   Procurement of food grains in the 1980s
      (according to crop year)   133
9.1   Some salient features of the Bhakra power plants   140
9.2   Hydro generating units under the BNP   141
9.3   Year-wise power generation   142
9.4   Transmission lines and sub-stations under the BNP   146
10.1 Month-wise Count of Annual Maximum Floods (1909–95)   153
10.2 Moderation of daily average flow by the Bhakra Reservoir
      during July–September, 1967–95   154
10.3 Moderation by the Bhakra dam   157
11.1 Poverty ratio   168
11.2 Milk Production in India   170
12.1 Sedimentation in the Bhakra Reservoir   187
12.2 Siltation in the Bhakra Reservoir   190
14.1 Growing electricity consumption in some countries   209
14.2 Per Capita Consumption of Electricity (kWh)   209

Figures

2.1   The Headworks Built on the Various Rivers of the
      Indus Basin During British Rule   17
2.2   Canals in the Indus Plains, as in 1947   18
3.1   Beas-Sutlej Link (General Layout)   34
3.2   Sutlej, Beas, Ravi rivers and Connected Main Canals   35
3.3   Projects Developed Under Indus Water Treaty   37
5.1   Index Map (Bhakra-Nangal Project)   70
5.2   Bhakra dam (Maximum Spillway Section)   71
9.1   BBMB Transmission Network   147

Boxes

Rally for the Valley   4
Partition   24
The millions who cannot vote   182
Swim out of ghost waters   212

GLOSSARY

Acre-foot	water needed to cover 1 acre to a depth of 1 foot (1233 cubic metres).
Abutment	part of valley side against which dam is constructed.
Aggradation	raising of riverbed due to sediment deposition.
Alluvium	sediments transported by a river and deposited on bed and floodplain.
Aquifer	geological formation of high porosity and high permeability that yields significant quantities of groundwater.
Barrage	a weir or other structure across the river equipped with sluice gates to head up and to regulate the water surface level in the river above the barrage and enable diversion to a canal.
Canal irrigation	irrigation with water supplied via canal, usually having been diverted from regulated river or reservoir.
Cofferdam	temporary dam built to keep riverbed dry to allow construction of permanent dam or other structure.
Culturable command area	that portion of the irrigable area which is commanded by flow irrigation.
Culturable irrigable area	the gross irrigable area less the area not available for cultivation.
Crown wasteland	uninhabited land belonging to government.
Consumptive use	non-reusable withdrawal of water; generally the quantity of water used by vegetative growth of crops.
Capacity of canal	the authorized full supply discharge.

Capacity factor	the ratio of the mean supply run in a canal during a crop season or year to its authorized full supply capacity.
Dependable supply	that quantum of river supply available in a sufficient number of years to be considered reliable for agricultural operations.
Dead storage	that part of a reservoir which is not used for operational purposes; the corresponding volume of water.
Delta	flat area of alluvium formed at the mouth of some rivers where the main stream divides into several distributaries before reaching sea.
Distributary	river branch flowing away from the main stream of a river; branch canal taking off from a main canal.
Drip irrigation	efficient irrigation system that delivers water directly to the roots of plants, for example, through perforated or porous pipes.
Earth (or earthfill) dam	a dam of compacted earth largely.
Embankment dam	dam constructed of excavated natural materials. Usually triangular in cross-section with a broad base that distributes weight over a wide area and can thus be constructed on a soft, unstable riverbed.
Evapo-transpiration	water transmitted to the atmosphere by a combination of evaporation from the soil and transpiration from plants.
Flood control	reducing flood risk by building dams and/or embankments, and/or altering river channel.
Flood management	reducing flood risk by actions such as discouraging floodplain development, establishing flood warning systems, protecting urban areas and isolated buildings, and allowing the most flood-prone areas to remain as wetlands.
Flood plain	area of valley inundated during floods.
Feeder	a canal which delivers water to another canal but may have no direct irrigation from it.

Gross area	the total area within the limits set for irrigation by a canal.
Gross irrigable area	the gross area less such areas within irrigation limits excluded for any reason.
Gigawatt	unit of power equal to 1,000 megawatts.
Gigawatt-hour	unit of energy equal to 1,000 megawatt-hours.
Gravity dam	dam built of concrete and/or masonry which relies on its own weight and internal strength for stability.
Groundwater	subsurface water contained in saturated soils and rocks.
Head	equals the vertical distance between the elevation of the surface of a reservoir and the surface of the river where water re-enters downstream.
Hyrdological cycle	the continuous interchange of water between land, sea or other water surface, and the atmosphere.
Headworks	the works constructed at the head of a canal, like a weir, barrage and the off-taking structure.
Irrigation efficiency	proportion of water used for crop growth relative to the total amount of water withdrawn by the irrigation system.
Intensity	the percentage of the CCA irrigated during a crop season or year.
Inundation canal	a canal dependent for its supply upon the natural surface level of water in a river.
Kharif	the summer crop or crop season (usually from April to September)
Kilowatt	unit of power equal to 1,000 watts.
Kilowatt-hour	unit of energy equal to 1,000 watt-hours.
Large-dam	defined by International Commission on Large Dams as a dam measuring 15 metres or more from foundation to crest. Dams of 10-15 metres may be defined as large dams if they meet certain special requirements.
Lift irrigation	usually refers to irrigation using groundwater but can include irrigation using water pumped from canals and reservoirs.

Live storage	that part of a reservoir that excludes dead storage; the corresponding volume of water.
Megawatt	unit of power equal to 1,000 kilowatts.
Megawatt-hour	unit of energy equal to 1,000 kilowatt-hours.
Non-perennial canal	in the Indian context, canal which runs only during kharif irrigation season.
Oustees	people displaced by development projects.
Perennial canal	a canal which runs throughout the year.
Peaking (peak) power	electricity supplied at times when demand is highest.
Penstock	pipe or shaft which delivers water to turbines.
Plant factor	relationship between a powerplant's capacity to generate electricity and the actual amount of electricity it generates.
Powerhouse	building or cavern housing turbines and generators.
Probable maximum flood (PMF)	largest flood considered reasonably possible at a site as a result of meteorological and hydrological conditions.
Rainwater harvesting	technique which conserves water by collecting rainwater run-off behind bunds or in small basins.
Reservoir capacity	the volume of water which can be stored in the reservoir.
Riparian	of or relating to or located on the bank of a river
Riverine ecosystem	zone of biological and environmental influence of a river and its floodplain.
Rockfill dam	embankment dam formed of compacted or dumped cobbles, boulders or rock fragments.
Run-of-river dam	dam which raises upstream water level but creates only a small reservoir and cannot effectively regulate downstream flows; barrage
Run-off	precipitation which drains into a watercourse rather than being absorbed
Rabi	the winter crop or season (usually from October to March)

Sailab	flood inundation
Salinization	the accumulation of salt in soil or water to a harmful level.
Sediment	mineral and organic matter transported or deposited by water or air.
Sediment load	amount of sediment carried by a river.
Sediment flushing	reservoir operation in which the reservoir is lowered so that fast-flowing water can erode accumulated sediments on the reservoir bed.
Selective withdrawals	water withdrawn from intakes at different reservoir elevations to influence thermal, physical and/or other properties of downstream water.
Silt	sediment composed of particles between 0.004 mm and 0.06 mm in diameter.
Sluice	structure with a gate for stopping or regulating flow of water.
Spillway	structure which discharges water from a reservoir.
Tailrace	channel by which water is discharged from the turbines into a river.
Tailwater	water in river immediately downstream from dam or tailrace.
Trap efficiency	the proportion of a river's total sediment load which is trapped in a reservoir.
Tube-well	deep, mechanically drilled well.
Water table	surface of groundwater.
Waterlogging	saturation of soil with water; a condition detrimental to plants.
Watt	unit of power equal to a rate of working of one joule per second.
Watt-hour	unit of energy equal to one watt supplied for one hour.

ACRONYMS AND ABBREVIATIONS

BBMB	Bhakra Beas Management Board
BCM	billion cubic metres
B/c Ratio	benefit over cost ratio
BNP	Bhakra-Nangal Project
CAD	Command Area Development
CADA	Command Area Development Authority
CAG	Comptroller and Auditor General
CBIP	Central Board of Irrigation and Power
CCA	Culturable Command Area
CEA	Central Electricity Authority
CGWB	Central Groundwater Board
CII	Confederation of Indian Industry
CPCB	Central Pollution Control Board
CPR	Centre for Policy Research
CSSRI	Central Soil Salinity Research Institute
Cumec	cubic metres per second
Cusec	cubic feet per second
CWC	Central Water Commission
CW&PC	Central Water & Power Connection
CWPRS	Central Water and Power Research Station
Delta	depth of irrigation measured at canal head
DPR	Detailed Project Report
DPAP	Drought-Prone Area Programme
D/s	downstream
DSM	Demand Side Management
EIA	environmental impact assessment
FAO	Food and Agriculture Organization
FICCI	Federation of Indian Chambers of Commerce & Industry
FRL	full reservoir level
FSL	full supply level
GO	Government Order

GDP	gross domestic product
GIA	gross irrigated area
GSA	gross sown area
GOI	Government of India
GW	groundwater
Ha	hectare
Ha m/Sq km/Year	hectare metre per square kilometre per year
HFL	high flood level
Hr	hour
HRD	human resource development
HYV	high yielding variety (ies)
ICID	International Commission on Irrigation and Drainage
ICOLD	International Commission on Large Dams
ICAR	Indian Council of Agricultural Research
IFPRI	International Food Policy Research Institute
IHA	International Hydropower Association
IRR	internal rate of return
ISWD	Inter State Water Dispute
IS	Indian Standards
ISI	Indian Standards Institution
IWT	Inland Water Transport
K	potassium
Kg	kilogram
Kg/ha	kilogram per hectare
KL	kilo litre
km	kilometre
km^2	square kilometre
km^3	cubic kilometre
kW	kilowatt
kWh	kilowatt-hour
l	litre
lPCD	litres per capita per day
LPS	litres per second
m	metre
m^3/d	cubic metres per day
Mcft	million cubic feet
MAF or M.Ac.ft	million acre feet
MAX	maximum
MCM	M cum *or* Mm^3 *or* million cubic metres

Mg	milligram
mg/l	milligram per litre
MGD	million gallons per day
M ha	million hectare
M ha m	million hectare metres
Min	minimum
MLD	million litres per day
mm	millimetre
MOA	Ministry of Agriculture
MOEF	Ministry of Environment and Forest
MOWR	Ministry of Water Resources
MT	million tonnes
MW	mega watt
N	nitrogen
NA	not available
NCA	net cropped area
NCT	National Capital Territory
NCIWRDP	National Commission for Integrated Water Resources Development Plan
NEERI	National Environmental Engineering Research Institute
NGO	non-governmental organization
NHPC	National Hydro Electric Power Corporation Ltd
NIA	net irrigated area
NRSA	National Remote Sensing Agency
NSS	National Sample Survey
NTPC	National Thermal Power Corporation
OECD	Organization for Economic Cooperation and Development
OFD	on farm development
O&M	operation and maintenance
P	phosphorous
PAC	Public Accounts Committee
PAPs	project-affected persons
PDS	Public Distribution System
PPM	parts per million
PSEB	Punjab State Electricity Board
RBA	Rashtriya Barh Ayog (National Flood Commission)
RM&U	renovation, modernization and uprating
R&D	research and development

R&R	resettlement and rehabilitation
Rs	rupees
RSEB	Rajasthan State Electricity Board
SAR	sodium absorption ratio
SSI	small-scale industries
STP	sewage treatment plant
STW	shallow tube-well
t	tonne
TAC	Technical Advisory Committee
TDs	total dissolved salts/total dissolved solids
TMC	thousand million cubic feet
TSS	total suspended solids
TVA	Tennessee Valley Authority
UN	United Nations
UNDP	United National Developemnt Programme
UNICEF	United Nations Children's Fund
u/s	upstream
USA	United States of America
USAID	United States Agency for International Development
USBR	United States Bureau of Reclamation
UT	Union Territory
WAPCOS	Water and Power Consultancy Services (India) Ltd
WCD	World Commission on Dams
WHO	World Health Organization
WMO	World Meteorological Organization

1

INTRODUCTION

Genesis

Worldwide, dams have played a key role in development. They were built to irrigate agricultural lands, supply water to meet domestic and industrial needs, generate power, control floods, and enable navigation. The remains of some of the earliest dams, dating back to at least 3000 BC, have been found in Yemen, Egypt, and China. The first use of dams for hydropower generation was around 1890. Since 1900, several large dams have come into existence. As technology advanced, the construction of larger dams and complex structures was undertaken.

Various countries are presently at different stages of economic development. Many affluent countries like Great Britain, the USA, and Canada, undertook dam construction and industrialization on a large scale from an early date. Of more than 8000 large dams in North and Central America, about four-fifths are in the USA. The Hoover dam, completed in 1936, remained the world's highest dam for more than two decades. Most of the affluent countries already have the necessary infrastructure for water resource management. Many developing and 'least developed' countries have been late entrants in dam construction activities and industrialization and still have a long way to go. As prosperity increased in the developed nations, some groups began arguing that the expected economic benefits from large dams were not being realized, and that environmental, economic, and social costs were unaccounted for. Others retorted that these armchair critics from rich countries, even while enjoying the lifestyle enabled by decades of dam construction in their part of the world, were opposed to the efforts of poor nations to improve their lot. By the mid-twentieth century, opposition to certain dams grew vocal and organized. Such debates gradually encompassed many dams in the undeveloped and developing countries, including India. Some publications[1]

---

[1] For instance, McCully, Patrick, *Silenced Rivers: The Ecology and Politics of Large Dams*, Zed Books, London, 1996.

were even brought out by activist NGOs trying to portray a bleak record of dams and the alleged political dimensions thereof.

Currently, proponents of dams point to their many benefits such as increased water availability to meet domestic and industrial needs, and increase in agricultural production through assured irrigation and generation of power. The opponents argue about the losses to the society and environmental costs, alleging that these outweigh the benefits. There has never been a real dialogue between these two camps. Instead, in recent decades, debates on large dams have become highly polarized and polemical, obfuscating the real issues. The bitter controversies regarding the Sardar Sarovar and Tehri dam projects in India are the most recent instances of such a polarized debate.

World Commission on Dams

The establishment of the World Commission on Dams (WCD) in early 1998, as also its Report, which was published in November 2000, have led to an even greater polarization of the 'dams debate', rendering it an acrimonious one rather than a balanced review.[2] The WCD Report questioned the very utility of dams and generated bitter debates regarding their impacts. Doubtless, many had a vested interest in this debate. The Report finally set out twenty-six guidelines for decision making. However, the greatest controversies and differences relate to these very guidelines.

A widely voiced criticism of the WCD was that it was constituted in a non-transparent manner, that the commissioners were chosen in an opaque manner, and that it was not a representative body of even the major stakeholders. Its Report highlighted the likely negative impact of dams, while remaining almost silent about their benefits. Wherever some contribution to development has been grudgingly admitted, it was invariably coupled with statements such as its being 'marred in many cases by significant environmental and social impact which, when viewed from today's values are unacceptable'. Much of its 'knowledge base' comprised submissions by interested parties and lobby forces, which had not been analysed for their accuracy. It also made selective use of the material from such a 'knowledge base'.

Reactions to the WCD Report have ranged from outright rejection to unabashed admiration. Immediately after the report was released, three

---

[2] WCD, *Dams and Development*, The Report of the World Commission on Dams, Earthscan Publications, London, 2000.

international professional bodies specializing in water resource development viz., the International Commission on Large Dams (ICOLD), the International Commission on Irrigation and Drainage (ICID), and the International Hydropower Association (IHA) wrote to the World Bank. They brought the main shortcomings of the WCD Report to the notice of the World Bank president. They stated, *inter alia,* that the WCD 'analysis of existing dams is unbalanced with a strong implication that the majority of the world's 40,000 dams are environmentally damaging or socially destructive.[3] Very little attention is devoted to the many well-known benefits of carefully planned dams and no feasible alternatives are suggested for meeting the future water, food and energy needs of the developing world'.

Many governments and NGOs expressed their disappointment and reservations on the Report, even as some showed their admiration. The International Rivers Network (IRN) promptly said, 'The WCD Report vindicates much of what dam critics have long argued'. The Government of India informed the Secretary General, WCD, on 1 February 2001 that the 'guidelines' for development suggested by the WCD 'are wholly incompatible' with 'India's development imperatives'. The letter concluded by stating that the recommendations and guidelines of the WCD 'are not acceptable'.

The World Bank, which was one of the sponsors of the WCD, itself had disagreements with the final outcome of the WCD Report. The Bank's Review noted that in the two years since the issuing of the WCD Report, no consensus has emerged on the applicability of the twenty-six WCD guidelines. The World Bank has its own guidelines, which differ in several important aspects from the WCD guidelines. Moreover, the World Bank has since reiterated its commitment to continued support in developing and managing priority hydraulic infrastructure in an environmentally and socially sustainable manner. The new strategy of the World Bank adopted in 2003 also recognizes that developing nations can reduce poverty only by making investments in infrastructure for development of water resources and that much of this is truly public infrastructure, where the Bank has a role to play.

Unsupported hyperbolic assertions were made by the anti-dam lobby and its supporters about 'not a single Big Dam' having delivered what it promised.[4] (See Box—RALLY FOR THE VALLEY)

[3] Joint letter from the Presidents of ICOLD, ICID, and IHA dated 28 December 2000 to James D. Wolfensohn, President, World Bank.

[4] A contingent of people were leaving Delhi on 29 July 1999 for a week's travel in the Narmada valley under the 'FREE THE NARMADA' campaign. The invitation letter included a note by Arundhati Roy about the impacts of big dams. The box item RALLY FOR THE VALLEY gives an excerpt from the invitation letter.

Excerpt from Arundhati Roy's note sent along with the invitation for

RALLY FOR THE VALLEY
FREE THE NARMADA
STOP THE DAMS

(29 July to 5 August 1999)

....

'INSIST ON HOPE'

For over half a century we've believed that Big Dams would deliver the people of India from hunger and poverty. The opposite has happened.

....

....

Not a single big dam in India has delivered what it promised. Not the power, not the irrigation, not the flood-control, not the drought-proofing. Instead, Big Dams have converted huge tracts of agricultural land into waterlogged salt wastelands, submerged hundreds of thousands of hectares of prime forest and pushed the country deep into debt.

The era of big dams is over. All over the world they are being recognized as technological disasters. As big mistakes. Yet in India, our Government refuses to review the situation.

....

ARUNDHATI ROY

The tendency to denigrate all engineering projects and engineers as exploitative and venal, and to blame the contractors who build, the engineers who plan and design, and the government which authorizes a dam project has become fashionable in India too. Presenting cases in a highly exaggerated manner in order to draw attention is also common. For instance, this is how one well-known critic of dams puts it—'The fact that they do more harm than good is no longer just conjecture. Big dams are obsolete. They're uncool. They're undemocratic. They are a government's way of accumulating authority.... They are a brazen means of taking water, land and irrigation away from the poor and gifting it to the rich'.[5]

[5] See Arundhati Roy, 'The Greater Common Good', *Outlook*, New Delhi, 24 May 1999, p. 56.

Dams are the means to an end, the planned end objective being economic and human development, leading to an improved lifestyle for the common people. Unlike countries in the temperate zone which receive rains round the year, India, under the influence of the monsoon, gets its annual rainfall in a limited period of three months, while water requirements are round the year. When fresh water resources are limited and very unevenly distributed over space and time, storage dams become necessary. India has, therefore, been building dams—small, medium, and large—for centuries. The pace of such activities accelerated after Independence from colonial rule. Increased agricultural production became inescapable for India because of rapid population expansion. Irrigation development on a greater scale became necessary to meet the requirements of the expanding population. The generation of hydropower, a relatively cleaner form of energy, also became necessary.

Performance Analysis

The WCD Report questioned the utility of dams and caused acrimonious debates on their impact. What got sidetracked were the scientific and rational evaluations of such projects and their performance evaluation over a long-enough period of service to enable appropriate conclusions to be drawn. There are hardly any objective *ex post facto* analyses of the various impacts of large dams. Unfortunately, there has also been little systematic collection of relevant data about dam projects in the past and in their absence, definite conclusions regarding their performance and impacts are difficult. Much data remain unavailable to the public. To be credible, post-project evaluations should be made through independent professional analyses, rather than leaving those to either the same agency that built and operated them, or to lobbyists. There have been precious few, if any, comprehensive independent analyses on how dam projects were selected for execution, how they performed over time, and on the returns we are getting from the investments. Instead, issues relating to dams, their benefits and impact, have become one of the battlegrounds in the sustainable development arena, as pointed out by Nelson Mandela while launching the WCD report.[6] He correctly noted that the 'problem, though, is not the dams. It is the hunger. It is the thirst. It is the darkness of a township. It is townships and rural huts without running water, lights, or sanitation'.

[6] Address by Nelson Mandela at London, 16 November 2000 on the occasion of the release of the WCD Report.

Many publications have been released dealing with the debate on dams and development. Most were prepared prior to the project completion and were predictions of their likely impact, rather than assessments of their actual performance. They give details from their own restricted perspectives and seek to draw conclusions but resolve very little. The WCD global review, which is the most recent study of this type, assembled a 'knowledge base', some based on studies and yet others based on submissions made before it. This provided the basis for the Commission's assessment of the technical, financial, economic, environmental, and social performance of large dams. The evaluation was based on the targets set for them by their proponents—the criteria that provided the basis for government approval.

In assessing the large dams reviewed by the WCD, the Commission found variability in delivering the predicted water and electricity services— and related social benefits—with a considerable portion falling short of physical and economic targets. It considered that large dams had a range of extensive consequences which were more negative than positive. They noted failures in implementing adequate mitigation, resettlement, and development programmes for the displaced. It stated that for most of the earlier projects, there was no reliable information on what happened to the displaced people.

The WCD concluded that the shortfalls in performance occurred and were compounded by significant social and environmental consequences, the costs of which were often disproportionately borne by the poor and other vulnerable groups. The Commission, however, pointedly stated that it was 'disturbed to find that substantial evaluations of completed projects are few in number, narrow in scope, poorly integrated across impact categories and scales and inadequately linked to decisions on operations.'

Whether one agrees with the WCD Report or not, there can be no denying the inadequacy of systematic post-project performance evaluations of large dams, globally or in India.

Study by the CPR

The Third World Centre for Water Management (TWCWM), assisted by the Nippon Foundation, Japan, initiated a project to assess how dams have contributed to the social and economic development of countries. The plan was to study the impact of selected dams in different countries, which have been operational for one or two decades. The Bhakra dam of the Bhakra-Nangal Project (BNP) in India was considered appropriate for such a study.

The objective was to closely study the BNP with regard to its positive and negative social, economic, and environmental consequences and to draw appropriate conclusions from this. The CPR, New Delhi, India, was entrusted with the study. The author, research professor of the CPR, who has been involved in water resources development and management for over five decades, served as the director of the project study, which was done during 2003–4.

The sphere of influence of the BNP extends over a vast area including a part or the whole of the states of Himachal Pradesh, Punjab, Haryana, Rajasthan, and the National Capital Territory of Delhi. The CPR study sought to ascertain all the 'standard' consequences of the BNP such as irrigation, electric power, flood management, increase in agricultural production, and drinking water availability in this region. The performance was evaluated against the promises made in the approved project report. The study sought to find the secondary and tertiary impact as well. These cover items such as increase in income, advances in health, education, transportation, and communication services, contributions to industrial and regional development, creation of a pool of trained personnel, and impact on women. Field visits were made to the project area and the affected regions. As the BNP was built and operated through the agencies of the government, much interaction with the central and state governments were needed. Fortunately, all these agencies offered their full cooperation to the study.

There has been no comprehensive evaluation of the performance of the BNP so far. In fact, no such up-to-date, holistic and systematic post-project performance evaluation of a large dam in India is available. However, many researchers, particularly economists, have used the BNP to test their economic studies or to illustrate their points of view. Such works done by K.N. Raj,[7] B.S. Minhas,[8] Kanchan Chopra,[9] and Ramesh Bhatia[10]—to cite the important ones among those brought to notice, were taken note of in the CPR study, even though none of them could be deemed to be holistic performance analyses of the BNP. The BBMB itself attempted such a study in 1988 and its work was fully taken into account.

---

[7] Raj, K.N., *Some Economic Aspects of the Bhakra Nangal Project*, Asia Publishing House, New Delhi, 1960.

[8] Minhas, B.S., Parikh, K.S., Marglin, S.A., and Weisskopf, T.E., *Scheduling the operation of the Bhakra system*, Statistical Publishing Society, Calcutta, July 1972.

[9] Chopra, Kanchan Ratna, *Dualism and Investment Patterns*, Tata McGraw-Hill, New Delhi, 1972.

[10] Bhatia, Ramesh, Scatasta, Monica, Cestti, Rita (eds), 'Indirect economic impacts of Dams', A study sponsored by the World Bank (forthcoming).

The perceptions of the people are most important for a study of the consequences of the project. Hence, interactions with representative sections of the people was an integral part of the CPR study. The field data collection from an extensive area, within the bounds of the limited budget available and the time frame, was undertaken between July 2003 and January 2004. The views of the affected people form part of the report.

This report is aimed at the readers, who are broadly aware of the picture of the economic development of India in general, and its water resources development in particular. It will serve those who seek to know more about some of the currently debated issues. As they might not be expert economists or engineers, the usage of technical jargon has been generally avoided, except to a minimum degree warranted in the circumstances. One should keep in mind that this presentation essentially deals with the BNP and any discussion beyond it is outside the scope of the study. For the same reason, no attempt has been made to write about the history of irrigation and power development of India through river valley projects or otherwise.

Peer Review

The report in its draft form was discussed in a workshop held on 12 January 2004 at New Delhi by a distinguished peer group of experts. A list of those who attended this workshop is given in Annexure-1. The Study Report was thereafter presented in May 2004, after carefully considering all the points and suggestions made by the peer group and others who were good enough to offer their views about the study.

Between May and October 2004, there were many other opportunities in India to critically examine the Bhakra-Nangal Study Report. An international workshop on the impact of large dams was also sponsored by the TWCWM, Nippon Foundation, and the IHA and supported by many other international professional organizations. This workshop took place in Istanbul, Turkey, in late October 2004. All suggestions and comments made there were taken into account and the Report was finalized for print in November 2004.

The Scheme of Presentation

This book has been arranged broadly as follows:

Chapters 2 and 3 discuss irrigation development in the Indus basin. This has been done in two parts. Chapter 2 delineates the development till partition of the country and Independence. The effects of partition on India in general and the BNP in particular are also discussed. Chapter 3 sets forth the developments in India in the post-Independence period, in the context of the Indus Treaty. Developments in Pakistan are also briefly traced.

Chapter 4 discusses the pre-project conditions in the project area and presents the available data. Any study of the economic, social, environmental, and other consequences of a water and land development project needs reliable baseline data to delineate the pre-project conditions in the relevant area. Unfortunately, such data is not always available, as there were no formal requirements to make impact assessments of a project till the 1980s. However, an attempt has been made to assemble in this chapter all such possible data in respect of the Punjab and Rajasthan areas of the pre-BNP years.

Chapter 5 gives details of the BNP, as was approved after Independence by the government. This chapter summarizes the salient features of the project and sets out the operational set-up as well as the regulatory provisions and guidelines for water releases.

Chapters 6 to 8 deal with the project's impact on irrigation. These chapters discuss the development of irrigated agriculture as enabled by the project, in the setting and ambience of the 'Green Revolution' that was simultaneously sweeping over Punjab, Haryana, and parts of Rajasthan and Uttar Pradesh. The period of study extends from the completion of the project construction till the full development of irrigation, unfettered by the constraints imposed by the Indus Treaty during the 'transition period' (till 1970) on the pace of development. Further progress till date has been covered too. The contribution to food production has also been examined, in the light of some controversial views on this subject.

Chapter 9 discusses the details of hydroelectric power and the impact of power generation through the project. The significant role played by the BNP in the operations of the Northern Electricity Grid has also been discussed.

Chapter 10 details the impact of the Bhakra reservoir on flood management in the flood plains of the Sutlej river downstream of the Bhakra and Nangal structures.

Chapter 11 deals with the many other incidental benefits of the project. The importance of the indirect and induced consequences has been pointed out.

Chapter 12 examines the silt load in the Sutlej river and the impact of the Bhakra dam by way of sedimentation in the reservoir. Its implications on the useful life of the project have also been discussed.

Chapter 13 indicates the perception of the people about the project's consequences based on a random survey of the impacted people. This encompassed different categories such as the displaced people from rural communities, the urban displaced, the irrigation benefited/affected, the impact of electrification, water supply, etc.

Chapter 14 seeks to sum up and identify some important lessons learnt from the performance of the BNP over the past five decades.

2

INDUS BASIN DEVELOPMENT TILL 1947

The Basin

The Indus, fed by the Himalayan snows and the rains of the Indian monsoon is one of the great rivers of the world. India gets her name from this very river. Its principal tributaries are the Kabul, joining it from the west, and the five rivers (from which the name 'Punjab' was derived) joining it from the east, namely, the Jhelum, the Chenab, the Ravi, the Beas, and the Sutlej.

The Indus rises in Tibet near the lofty mountains around Manasarovar. Its course from the Himalayas to the sea is over 3000 km long [with over 1100 km in Jammu and Kashmir (J&K)]. The Indus basin is bounded by the Himalayas in the east, by the Karakoram and Haramosh ranges in the north, by the Suleimanki and Kirthar ranges in the west, and by the Arabian Sea in the south. The system drains an area of 1.15 million sq. km. The drainage area of the Indus basin in India is 321,289 sq. km.

The Indus enters the state of J&K in India at an elevation of over 4200 metres. In its run, the river skirts Leh and crosses the trade route to Central Asia via the Karakoram pass. Beyond Skardu, it crosses the Haramosh mountain, where it takes an acute turn southward and enters Kohistan. At about 1450 km from its source, the Indus is joined by the Kabul and continues to flow southwards. Eight hundred kilometres upstream of the sea, the Indus receives the Panjnad, the accumulated waters of its five eastern tributaries.

The Jhelum, known as the Vitasta, Vyeth, or Veth in J&K, has its source in Verinag at the upper end of the Kashmir valley. Below Srinagar, it receives the Sind and reaches the Wular lake. Below Baramula, the river rushes over a deep gorge between lofty mountains. At Muzaffarabad, the Kishanganga joins it. Lower down, it debouches into the plains near the city of Jhelum in Pakistan.

The Chenab rises in two headstreams, the Chandra and the Bhaga, in Himachal Pradesh. The united river is then called the Chandrabhaga or the

Chenab. It flows through the Pangi valley and enters Kashmir at an elevation of over 1800 metres. It then flows for over 290 km between steep cliffs, and then for 40 km through the lower hills to Akhnoor, where it enters Pakistan near the Marala weir, below the junction of the Tawi. In Pakistan, it flows for some 640 km to Panjnad, where it joins the Sutlej.

The Ravi rises near the Rohtang pass in Himachal Pradesh. After crossing the Siwaliks, it enters the Punjab plains near Madhopur. It passes into Pakistan some 26 km below Amritsar.

The Beas, too, rises on the southern face of the Rohtang pass at some 4060 metres above sea level. It reaches the Siwaliks near Hoshiarpur and finally joins the Sutlej near Harike, after a total course of 460 km—wholly in India.

The Sutlej rises in Tibet near the Manasarovar lake and has a long course through the mountain ranges, rising to 6000 metres on either side. After its course through Himachal Pradesh, it enters Punjab in the Hoshiarpur district and emerges from the Siwalik hills at the Bhakra gorge. It receives the Beas at Harike above Ferozpur and finally joins the Chenab at Madwala in Pakistan.

The mean annual flows in the various rivers in the Indus system have been indicated in Table 2.1.

Table 2.1: Average annual flows in the rivers of the Indus system

River	Mean annual flow MAF	Mean annual flow BCM
Indus	61.9	76.4
Kabul	27.6	34.0
Jhelum	22.6	27.9
Chenab	23.5	29.0
Sub-total—'western rivers'	135.6	167.3
Ravi	6.4	7.9
Beas	12.8	15.8
Sutlej	13.6	16.8
Sub-total—'eastern rivers'	32.8	40.5
Total Indus system	168.4	207.8

Source: *Records of the Indus Treaty negotiations and published documents thereon.*
Note: The flows are as measured at the rim stations, where they emerge from the Himalayan foothills. They are based on 25 years' data (1921–2 to 1945–6).

A special feature of the rivers of the Indus system is that they get their waters only from the upper mountainous catchments and carry their highest flows while emerging from the foothills. The additions to the flow from the large but arid plains are insignificant. There are also variations in the total river flows from year to year.

The Indus basin is one of the world's fertile but arid areas. The plains of the Indus receive low and undependable rainfall, thus making them one of the arid regions of the world. The annual rainfall in the Indus plains decreases from northeast to southwest, from about 75 cm in the upper part to 25 cm lower down. Moreover, a large part of the annual rainfall occurs during the three monsoon months of July to September. Consequently, in large parts of the Indus basin, no cultivation is possible without irrigation. No other comparable area in the world is as well adapted to irrigation as the flat and arid plains of the Indus basin, with six perennial rivers flowing through them.[1]

Early Developments

Irrigation development in the Indus basin goes back to ancient times as far as the Harappan Civilization (or Indus Valley Civilization), when small groups of people, whose main source of subsistence was agriculture, settled down as communities along the river banks. The annual inundation of lands when the river was in flood, coupled with whatever rainfall occurred, provided enough moisture for raising crops. This type of irrigation known as *Sailab* appears to have been the main form of irrigation in those days.

Well irrigation has also been traced to ancient times in the Indus valley, where water-bearing strata of good quality were found in abundance. Different types of water-lifting devices were used as well.

The existence of an inundation canal system in the Indus basin is recorded at the time of the Greek invasion of India in 326 BC. During the Mauryan rule (321 BC to 185 BC), public irrigation works, based both on river and tanks, helped farm production. The *Arthasastra* of Kautilya (third century BC) gives details of modes and rules of irrigation practised during that period.

In the Vitasta (Jhelum) valley in Kashmir, extensive engineering operations were carried out during the reign of King Avantivarman, for its drainage and irrigation system. The systematic regulation of the river course by engineer Suyya reduced flood damages and created additional land for cultivation and irrigation (ninth century AD).[2]

---

[1] Gulhati, N.D., 'The Indus and its Tributaries', in B.C. Law (ed.), *Mountains and Rivers of India*, 21st International Geographical Congress India, 1968, p. 352.

[2] Kalhana's *Rajatarangini* (M.A. Stein's translation and commentaries), Motilal Banarasidass Publishers, Delhi, 1900 (reprinted 1989).

After the Islamic invasions of India commenced, the Delhi Sultanate period is mentioned as one of importance for the steps taken to rehabilitate agriculture. Firoz Shah Tughlak (AD 1351–88) was reportedly a pioneer in the history of canal irrigation. In about AD 1355, he built a canal to carry the Yamuna waters to his hunting grounds in the Hisar district. During the Mughal rule (AD 1526 to AD 1819), some irrigation and water supply works were carried out. In Emperor Akbar's reign, his son Mohammed Salim asked that the canal to Hisar be renovated. The first perennial canal in the Indus basin was reportedly built in Jahangir's period (AD 1605 to AD 1627). His successor, Shah Jahan (AD 1628–58) built a canal that emerged from the Ravi.

R.B. Buckley has described the system of inundation canals built by the Mughal rulers and their successors as they existed in AD 1872.[3] According to him, the area lying between the Sutlej and the Chenab was rendered fertile, despite very little rainfall, by a series of inundation canals taken from the rivers. These were originally constructed during the reign of Aurangzeb (AD 1658 to 1707). Similarly, there was a group of twelve inundation canals on the right bank of the Indus above Mithankot, and another group which irrigated a tract about 20 km wide on the left bank in the Thal Doab. There were also inundation canals irrigating areas lying between the Sutlej and the Ravi, besides those in the Bahawalpur area that emerged from the Sutlej river at Panjnad. Thus, at the beginning of their rule in India, the British inherited a network of inundation canal systems, functioning to different degrees of efficiency in different areas. Most of those canals, however, had gradually become silted. By 1850, only some inundation canals remained to serve the low-lying lands near the rivers, which were inundated by rising flood water in the summer. Hence, the first task of the British engineers was to attend to these canals with a view to improving their performance and extending their scope by bringing more areas under irrigation. As these canals were without any headworks, the engineers had to develop a secure and streamlined system of diverting the river water into them.

Developments Under British Rule in India

The British entered the Indus basin area in 1819. After the British annexed Punjab in 1849, they considered it politically necessary to urgently settle the large number of disbanded Sikh soldiers. For this purpose, it was decided to construct a canal that began from the Ravi river to irrigate the

---

[3] Buckley, R.B., 1893, *Irrigation Works in India and Egypt*, Encyclopaedia Britannica, London, 1970.

arid tracts of Central Punjab. Thus was born the Upper Bari Doab canal (UBDC). When originally conceived and constructed between 1851 and 1859, it was known as the Bari-Doab canal, which had to be renamed as the UBDC when work on a Lower Bari-Doab canal was taken up in 1907. The original idea was to utilize, to the extent possible, the old Hansli canal built in Shah Jahan's time. However, eventually, based on systematic surveys, it was decided that the old canal could not be used and, hence, a new one was needed.

When the UBDC was opened in 1859, there were no permanent head-works nor were any distributary channels constructed. Wooden crates filled with boulders served as a weir but such work had to be undertaken every year after the flood season. In 1868, the construction of the permanent head-works at Madhopur was taken up. However, as this got damaged during construction, the scheme was revised in 1874 and its scope enlarged by including a properly designed distribution system. The revised scheme was completed in 1879.

The Sirhind canal project was undertaken in 1873 and irrigation com-menced in 1882. This project was financed from public loan funds and was built by a governmental agency to irrigate areas not only in British-controlled Punjab but also the princely states of Patiala, Nabha, and Jind, without considering political boundaries. It necessitated the preparation of agreements between the Punjab Government and the princely states.

The Sirhind canal begins from the Sutlej river at Rupar, at a point which is 80 km (50 miles) downstream of Bhakra. The original headworks of the canal consisted of a weir. This eventually had to be remodelled, however, as it was defective and induced heavy silting in the canal. Subsequently, the weir was fitted with 1.8 m (6 ft) high falling shutters, the waterway increased and a silt trap provided by way of the still pond method in front of the regulator.

The lower Sohag and Para canals that began from the Sutlej and were built in 1882, and the Sidhnai canal from the Ravi that was built in 1886, provided irrigation to areas that were hitherto uncultivated wastelands owned by the government (called 'crown waste'). Encouraged by their profits, the government sanctioned the Lower Chenab canal and the Lower Jhelum canal in 1890. These projects marked the beginning of 'canal colonies'. As these wastelands had no resident population, in addition to the provision of irrigation canals, whole communities were also brought into the area. They soon converted the wilderness into prosperous lands.

In accordance with the recommendations of the First Irrigation Commis-sion (1901–3), several projects were taken up in the early twentieth century.

These included the Triple Canal Project in West Punjab, which was a landmark in Indian irrigation as it linked three big rivers—the Jhelum, the Chenab, and the Ravi. The Upper Jhelum canal beginning at Mangla on the Jhelum, the Upper Chenab canal starting at Marala on the Chenab, and the Lower Bari Doab canal beginning at Bulloki on the Ravi, comprised the 'Triple Canals'.

Soon after the end of the First World War, a number of new schemes were initiated for developing irrigation in different parts of the Indus basin. These included the Sutlej Valley Project—a system of three headworks and nine canals from the Sutlej river, Panjnad headworks on the Panjnad with two canals, and the Sukkur barrage across the Indus with seven canals. All these were completed by 1935. Later, the Haveli and Thal canals were sanctioned in 1937 and completed before Independence.

Irrigators in the Indus basin had got used to varying and unpredictable canal supplies, both within a crop season as also from year to year, and tried to make the best of the situation. To ensure that each part of the canal system received an equitable share of the available waters, rotation of supplies to distributaries with proper maintenance of water accounts were resorted to. Over decades of such management, a degree of efficiency was achieved. The impetus this experience gave to the evolution of modern irrigation engineering received worldwide recognition.

The pre-1947 canal systems included some of the biggest irrigation achievements in the world, such as the Triple Canals Project and the Sukkur barrage. The most remarkable feature is that the entire pre-1947 irrigation development in the Indus basin, covering 26 million acres (10.5 million hectares), was brought about without the construction of *any* storage reservoir. In 1947, this area was larger than the total irrigated area in any other country of the world, except China. The USA, the next most irrigated country then, irrigated 23 million acres (9.3 million hectares). About 73 MAF (90 BCM) of water was being annually diverted (in 1945–6) from the rivers of the Indus system for the irrigation canals irrigating these 26 million acres. The water that flowed annually to the Arabian Sea was assessed at 83 MAF (102 BCM), on the basis of the average flows for the five years 1941–2 to 1945–6. This is, of course, the average picture, with annual and seasonal variations. In addition, there were a large number of wells, mostly animal-powered shallow percolation ones, that irrigated a further 4 to 5 million acres (about 2 million hectares) annually.

Various headworks built on the different rivers of the Indus basin under the British rule, as of 1947, are shown in Figure 2.1. The main canals in the Indus plains taking off from these headworks are shown in Figure 2.2.

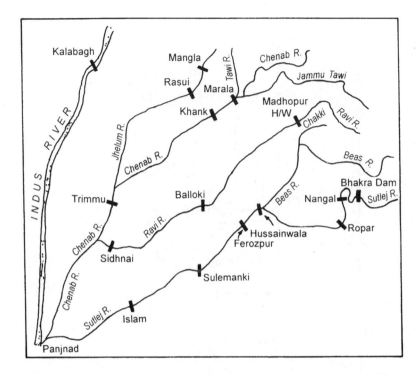

Figure 2.1: The Headworks Built on the Various Rivers of the Indus Basin During British Rule

Bhakra—The 'Site Made by God for Storage'

The BNP derives its name from two small villages in the state of Punjab. There are two integral parts of this multi-purpose river valley project, namely, the 'Bhakra dam' and the 'Nangal barrage'. The first part of the scheme is a storage dam at Bhakra across the Sutlej river, the major tributary of the Indus river system. The second is a barrage at Nangal, 13 km (8 miles) downstream of Bhakra village on the Sutlej, which enables the diversion of water to the Nangal hydel canal and feeding the Bhakra irrigation canal system.

The origins of the Bhakra dam can be traced to Sir Louis Dane, a former Lieutenant Governor of Punjab. In November 1908, he was floating down the Sutlej river after a visit to the princely state of Bilaspur. On the way, Sir Dane, who was trained as an engineer before being inducted in the Indian

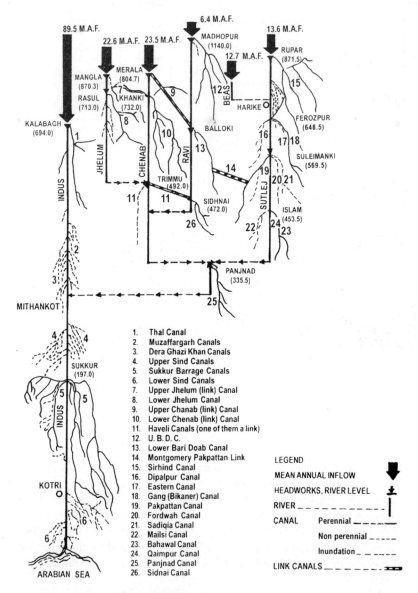

Figure 2.2: Canals in the Indus Plains, as in 1947

Civil Service, noticed the narrow gorge near Bhakra with high abutments on either bank. It occurred to him that 'here was a site made by God for storage'.[4] In a note dated 18 November 1908, he asked the Irrigation Department of the Punjab province to explore this possibility. Mr Gordon, Chief Engineer, Punjab Irrigation, after a visit to the site in November 1909, considered a dam at Neilla (some 5 km downstream of the finally chosen site at Bhakra) as most promising. Mr Baker of the Punjab Irrigation Department prepared a detailed report. However, in 1910, the project cost was considered prohibitive when compared to the benefits and further investigations were stopped.

Chief Engineer F.E. Gwyther revived the proposal in 1915, as he considered that the benefits and potentials for earning revenues had been underestimated and were needlessly pessimistic. After detailed investigations, the first detailed project report was prepared in 1919 by Er H.W. Nicholson. It envisaged a dam at Bhakra, 121 m (395 feet) high, to store 3.18 BCM (2.58 MAF) of water to extend irrigation to the traditionally famine-affected areas of the province like Hisar and Rohtak, as also the adjoining princely states of Patiala, Jind, Faridkot, and Bikaner. The project was not accepted. Instead, the government sanctioned the Sutlej Valley Project.

Although the project was not taken up for construction, further investigations and the examination of the Bhakra gorge continued. Geological studies were undertaken too. In 1927, an expert committee was formed to report on the dam site and other storage possibilities between the Yamuna and the Chenab. The committee visited Bhakra in November 1927. Its report recommended a higher storage facility, with a 152.4 m (500 ft) high dam at the gorge. Accordingly, the Punjab government revised the earlier scheme in 1928. Reservoir surveys were undertaken in 1932. This scheme, too, did not fructify, as the colonial government of the time had its own priorities that envisaged development of 'crown waste' lands in western Punjab. Soon, many elected members of the legislature, representing areas in east Punjab, pressed for undertaking the Bhakra scheme. Sir Chhotu Ram, who became revenue minister in Punjab in the early 1940s (he was originally from Rohtak town in the famine-stricken area of south-east Punjab) felt that the Bhakra dam project was the first essential to help banish famine in that area. He relentlessly pursued the subject till his death in January 1945. The project proposals were revived once again in the early 1940s and Dr A.N. Khosla put in charge of the investigations. The project now envisaged the full

---

[4] Sain, Kanwar, *Reminiscences of an Engineer*, Young Asia Publications, Ne⸳. Delhi, 1978, p. 58.

reservoir level at 487.68 m (1600 ft) providing a storage of 5.86 BCM (4.75 MAF). Hydroelectric power generation and irrigation benefits were envisaged.

In 1944, the Punjab government invited Dr J.L. Savage, Chief Engineer, US Bureau of Reclamation, to examine the site and report on the feasibility of the dam. He observed that the site was suitable and recommended some further explorations. The American geologist, F.A. Nickel, came to help the Punjab engineers with these investigations. These and related studies suggested an additional raising of the dam height, subject to certain conditions. However, the existence of the princely state of Bilaspur upstream of the proposed dam site and the Raja's reservation over submergence within his territory was another factor that needed to be taken into account. In 1941–2, Khosla, who was then the superintending engineer succeeded in securing the Raja of Bilaspur's agreement to the construction of the dam at Bhakra and to the submergence of large areas in his state. The terms of this agreement were accepted in February 1945 and formalized later that year. Specification designs for a straight gravity concrete dam, then under preparation by the International Engineering Company, USA, were, therefore, asked to be made with elevation 481.58 m (1580 ft) as the reservoir level.

During the years of the Second World War, Punjab suffered an acute power shortage and the Nangal hydel scheme was mooted as the immediate solution. As a result, while negotiations continued and preliminary works for the Bhakra dam were undertaken, it was considered desirable to give priority to the Nangal barrage, the hydel canal, and the irrigation canal system.

Partition of India

Major political changes soon swept over the subcontinent. Britain was preparing to transfer political power over India.[5] Prime Minister Clement Attlee made a policy statement on 20 February 1947 in the House of Commons that transfer of power 'into responsible Indian hands by a date

---

[5] The twelve volumes of official documents relating to the transfer of power to India were published by Great Britain titled *Constitutional Relations between Britain and India: The Transfer of Power 1942–47*, Chief Editor Dr Nicholas Mansergh (1970–83). Many other versions are available which look at the picture from different viewpoints. Many eyewitness accounts too were published. H.M. Seervai (1994) has made an interesting presentation relating to Partition titled *Partition of India: Legend and Reality*.

not later than June 1948' would take place. A new Viceroy, Lord Mountbatten, was named to replace Lord Wavell and directed to implement the transfer of power. Ultimately, some parts of British India with 'contiguous majority areas of Muslims' were carved out to form the new state of Pakistan.[6] The Punjab province, too, got divided and the western part went to Pakistan. The Partition of India, as also of Punjab, had serious implications for the BNP.

It would be outside the scope of the present work to enter into a detailed study and analysis of the circumstances and the manner of carving out a part of India to form a new nation as a part of achieving India's Independence from British rule. However, it is necessary to take note of the facts relating to the Partition of the Punjab, which have a deep bearing on the subject of this study. The 'powers that be' rushed through the entire process of Partition in such a short time that even half a century later, its consequences are still being unravelled. Lord Mountbatten was sworn in as the Viceroy on 24 March 1947, and immediately started political discussions on the transfer of power.[7] All major political parties quickly accepted the 'plan for the immediate transfer of power' and the statement of the government to this effect was made on 3 June 1947. Thereafter, nothing mattered more than speed for the viceroy. The British Parliament passed the Indian Independence Act and royal assent was given on 18 July 1947. By the end of June, the provincial assemblies of Bengal and Punjab had voted for the partition of their provinces.[8] The British government had already selected the eminent lawyer, Sir Cyril Radcliffe, and recommended that he might head the Boundary Commission. Lord Mountbatten quickly obtained the concurrence of the two major political parties, the Congress and the Muslim League, for the proposal to invite Radcliffe as chairman of both the Boundary Commissions. On 30 June, the Government of India announced the setting up of two Boundary Commissions, one each for Bengal and Punjab, further indicating that the same person would be the chairman of these commissions. The Punjab Boundary Commission was 'instructed to demarcate the boundaries of the two parts of the Punjab on the basis of ascertaining the contiguous majority areas of Muslims and non-Muslims. In doing so it will also take into account other factors'.[9] Radcliffe was in India by 8 July and within forty days his awards were ready. They

---

[6] There are many official accounts of the partition of India. The immediate and basic official account was published by the Partition Secretariat in 1950.

[7] Ziegler, Philip, *Mountbatten*, Harper & Row, New York, 1985, p. 367.

[8] Ibid., p. 367.

[9] Government of India, *Gazette Extraordinary*, 30 June 1947.

were made public on 17 August 1947. When Pakistan came into existence on 14 August 1947 and when India celebrated its Independence, the public was unaware of the exact boundaries between Pakistan and India.

When India emerged as a free nation from its colonial past, it suddenly faced not only the vivisection of its territory but also the cutting up of its rivers and canal systems. Sixteen districts of the Punjab province went over to West Pakistan, along with some other provinces. Thirteen others, forming East Punjab remained with India. The Independence Act did not concern itself with the use of the waters of the Indus system by the successor states, nor was it possible to do so, as the boundary was then not known. Even Sir Cyril Radcliffe threw up his hands in despair on this matter. The boundary line drawn by the 'Radcliffe award' overnight cut across the Punjab state harshly. Of the then existing thirteen canal systems of the Indus basin, ten went to Pakistan, two remained with India and one (the UBDC) was divided between the two countries. Many of the headworks feeding the systems lay in different countries. In some cases, while the headworks were in one country, most of the protective embankments were in another. Sir Penderel Moon, who was then a minister in Bahawalpur state, commented that, 'It seemed extraordinary that there had been no one to impress upon Lord Radcliffe the importance of including the principal protective works in the same territory as the head works. This could have been easily done'.[10]

A Nugget of Doubt About Ferozpur

It is a matter of record that the political parties were very concerned about the canal systems and headworks on the rivers of the Indus system, particularly on the Sutlej. They made representations on the manner in which these should be kept functional in the future, or how the boundary line being demarcated would affect them and the future interests of India and Pakistan. The Boundary Commission itself had worries in this regard. It seems that, at some stage, the commissioners were giving thought to joint control of works and the likely effects thereof. In his award, Radcliffe expressed the hope that where the new boundary 'cannot avoid disrupting such unitary services as canal irrigation, railways and electric power transmission, a solution may be found by agreement between the two states for some joint control of what has hitherto been a valuable common service'.

[10] Moon, Penderel, *Divide and Quit*, Oxford University Press, New Delhi, 1998, p. 186.

Around 8 August 1947, it became known that Radcliffe was likely to award the Ferozpur and Zira *tehsils* to Pakistan. Khosla considered that this would be disastrous from the viewpoint of East Punjab and Bikaner state and brought it to the notice of Nehru. He was also against any joint control of irrigation canals. Nehru thought that Khosla's views were valuable and passed on his advice to Mountbatten on 9 August, so that the latter could consider appraising Lord Radcliffe. The Maharaja of Bikaner had also moved similarly on 10 August. Mountbatten formally informed all of them that the viceroy had nothing to do with the Boundary Commission and that he had stayed clear of the entire issue.

However, many independent accounts have come to light since then, which allege that on 8 August, Radcliffe had indeed completed the Punjab line allotting the Ferozpur and Zira sub-districts to Pakistan. After his lunch with Mountbatten two or three days later, Radcliffe altered the line by allotting Ferozpur and Zira to India. Philip Ziegler has referred to this issue in his biography of Mountbatten. He then defends Mountbatten in the following words:[11] 'To argue that Mountbatten tampered with the awards is to suggest that Radcliffe, a man of monumental integrity and independence of mind, meekly allowed his recommendations to be set aside by somebody who had no official standing in the matter.' However, he added later, 'Yet a nugget of uncertainty remains', and refers to many other circumstantial evidence which go to support that feeling of uncertainty.

Volume XII of *Transfer of Power* dealing with this period of history was published around the same time as Ziegler's biography. Many documents included in this official version lend credence to the doubts expressed over Mountbatten's role in the matter. Moreover, on 24 February 1992 the *Daily Telegraph* (London) carried a statement by Mr Beaumont (personal secretary to Radcliffe) to the effect that 'Radcliffe was persuaded to change his mind about Ferozpur and Zira at a lunch with Mountbatten', from which Beaumont was 'deftly excluded'. H.M. Seervai, who had been pursuing the matter for long, in his book *Partition of India*, agrees with the version that Mountbatten indeed had played a role in Radcliffe's change of mind.[12]

---

[11] Ziegler, Philip, *Mountbatten*, Harper & Row, New York, 1985, p. 421.

[12] Seervai, H.M., *Partition of India: Legend and Reality*, N.M. Tripathy Publishers, Bombay, Second edition, 1994.

Partition

Unbiased at least he was when he arrived on his mission,
Having never set his eyes on this land he was called to Partition
Between two peoples fanatically at odds.
With their different diets and incompatible gods.
'Time', they had briefed him in London, 'is short. It's too late
For mutual reconciliation or rational debate:
The only solution now lies in separation.
The Viceroy thinks, as you will see from his letter,
That the less you are seen in his company the better,
So we've arranged to provide you with other accommodation.
We can give you four judges, two Moslems and two Hindu,
To consult with, but the final decision must rest with you'.
Shut up in a lonely mansion, with police night and day
Patrolling the gardens to keep assassins away.
He got down to work, to the task of settling the fate
Of millions. The maps at his disposal were out of date.
And the Census returns almost certainly incorrect,
But there was no time to check them, no time to inspect
Contested areas. The weather was frightfully hot,
And a bout of dysentery kept him constantly on the trot,
But in seven weeks it was done, the frontiers decided,
A continent, for better or worse divided.
The next day he sailed for England, where he quickly forgot
The case, as a good lawyer must. Return he would not,
Afraid, as he told his club, that he might get shot.

W. H. Auden

3

INDUS BASIN DEVELOPMENT AFTER 1947

At Independence

All the important rivers of the Indus system, including the main river, originate in India, or pass through India before crossing over to Pakistan. A water dispute between the two parts of Punjab on the use and control of the waters of the Indus system arose overnight. This soon enlarged into an international water dispute which complicated the already bitter relations between India and Pakistan. The Maharaja of J&K had acceded to India by signing the Instrument of Accession in October 1947, as per the India Independence Act. Therefore, the occupation of certain areas of J&K by Pakistan made matters worse.

Forty-six million people were living in the Indus basin prior to Partition: 21 million in the part that came to independent India and 25 million in the area that went to Pakistan. The culturable irrigable land in the basin was 65 million acres, of which 26 million acres remained in India and 39 million acres went to Pakistan. Out of the 28 million acres under irrigation at the time of Partition, only 5 million acres were left in India. In other words, of the 26 million acres of cultivable land of the basin that came to India, only 5 million acres (that is less than one-fifth) were irrigated land. These 28 million acres of irrigated land were using some 75 MAF of the Indus system waters, of which, as of 1947, about 66 MAF were used in the Pakistan canals and another 9 MAF in the Indian canals.

Post-Independence Developments in India

It will be difficult for the people to forget the incalculable damage caused by the Radcliffe Award. The Partition holocaust led to the sudden uprooting of millions of people from their hearths and homes in this region.[1]

---

[1] Various estimates exist of the number of displaced and the exact figures are not known. The figures vary from 12 to 17 million people. The casualties were put at half to 1 million people.

Practically the entire Hindu-Sikh population of West Pakistan migrated to India and most Muslims of East Punjab went over to West Pakistan. This sudden mass migration of millions of people and their rehabilitation in India necessitated one of the biggest and toughest land settlement operations in world history. These refugees had to begin their lives afresh in India. Among those who came over to India from West Pakistan were the industrious cultivators who had developed the canal colonies of West Punjab and Sind, and had gained vast experience in converting arid wastelands to rich cultivated fields. There were also industrialists— big, medium, and small—as well as skilled workers who had lost their workplaces and jobs overnight. These people needed land for cultivation, and power for setting up industries, and opportunities to restart their livelihoods. The BNP was considered the answer to meet these needs.

HARIKE BARRAGE

After Independence, in addition to extending irrigation to new areas in India, the immediate need arose to improve the headworks across the Sutlej river near Ferozpur, (at the new Indo-Pakistan border). Therefore, the construction of the Harike barrage, at a point about 3 km downstream of the confluence of the Sutlej and the Beas, was taken up. The work was completed in 1950. Certain areas in the tail portion of the Sirhind canal were, thereafter, shifted to receive irrigation supplies from the Harike barrage. The Makhu canal, the Ferozpur feeder, and the Rajasthan feeder begin from the Harike barrage.

BHAKRA DAM

The major Sutlej river developments after Independence comprise the BNP and improvements in the Sirhind canal. Soon after Partition, to facilitate the finalization and early start of work on the Bhakra dam, Khosla undertook active discussions with the Raja of Bilaspur on the project. He was successful in persuading the Raja of Bilaspur to accept a higher dam under the changed circumstances after Independence, both in the larger interest of the country as a whole, as well as of the Bilaspur state. An understanding was reached that a modern township would be constructed to replace the then existing Bilaspur town that would be submerged by the higher dam. This understanding was reached even before the later merger of the state with the Indian Union. It was decided to construct the dam to the maximum safe height—as determined by the foundation conditions— and to fully exploit the irrigation and power potential of the Sutlej. This decision of 1948 meant that the storage level would go up to 1680 ft elevation and the top of the dam to 1700 ft, a 100 ft higher than envisaged earlier.

A WATER DISPUTE BEGINS

By early 1948, serious differences arose between India and Pakistan on the use of the Indus waters and on various proposals for the future, including the Bhakra project. Pakistan raised objections to the project. West Punjab started digging a new supply channel from the Sutlej, bypassing the Ferozpur headworks in India and endangering its safety.

At the instance of West Punjab, the chief engineers of East Punjab and West Punjab entered into an agreement on 20 December 1947, to continue the *status quo ante* on the UBDC and at Ferozpur (Sutlej valley canals). The Punjab Partition Committee approved it the same day. This agreement was to end on 31 March 1948, before which date a further arrangement for any period beyond could be negotiated. However, West Punjab did not take any steps till the end of March for an agreement for any further period. East Punjab discontinued the releases for the UBDC on 1 April 1948, with the result that some irrigation channels near Lahore went dry. Prime Minister Liaqat Ali Khan of Pakistan, in a telegram addressed to the Indian Prime Minister Jawaharlal Nehru, regretted the stoppage of water and requested that it be restored immediately.

The chief engineers of East and West Punjab met on 15 April 1948 and reached an agreement to be effective from the date of ratification by the two Dominions of India and Pakistan. Prime Minister Nehru ordered the immediate resumption of water supply and fixed 3 May 1948 for the meeting at the government level. On 4 May 1948, an agreement was reached. It is reproduced below because of its importance.

1. A dispute has arisen between the East and West Punjab Governments regarding the supply by East Punjab of water to the Central Bari Doab and Depalpur canals in West Punjab. The contention of the East Punjab Government is that under the Punjab Partition (Apportionment of Assets and Liabilities) order, 1947, and the Arbitral Award the proprietary rights in the waters of the river in East Punjab vest wholly in the East Punjab Government and that the West Punjab Government cannot claim any share of these waters as a right. The West Punjab Government disputes this contention, its view being that the point has conclusively been decided in its favour by implication by the Arbitral Award and that in accordance with international law and equity, West Punjab has a right to the waters of the East Punjab rivers.

2. The East Punjab Government has revived the flow of water into these canals on certain conditions of which two are disputed by West Punjab. One, which arises out of the contention in paragraph 1, is the right to the levy of seigniorage charges for water and the other is the question of the capital cost of the Madhavpur Headworks and carrier channels to be taken into account.

3. The East and West Punjab Governments are anxious that this question should be settled in a spirit of goodwill and friendship. Without prejudice to its legal rights in the matter, the East Punjab Government has assured the West Punjab Government that it has no intention suddenly to withhold water from West Punjab without giving it time to tap alternative sources. The West Punjab Government on its part recognize the natural anxiety of the East Punjab Government to discharge the obligation to develop areas where water is scarce and which were underdeveloped in relation to parts of West Punjab.

4. Apart, therefore, from the question of law involved, the Governments are anxious to approach the problem in a practical spirit on the basis of the East Punjab Government progressively diminishing its supply to these canals in order to give reasonable time to enable the West Punjab Government to tap alternate sources.

5. The West Punjab Government has agreed to deposit immediately in the Reserve Bank such ad hoc sum as may be specified by the Prime Minister of India. Out of this sum, that Government agrees to the immediate transfer to East Punjab Government of sums over which there is no dispute.

6. After an examination by each party of the legal issues, of the method of estimating the cost of water to be supplied by the East Punjab Government and of the technical survey of water resources and the means of using them for supply to these canals, the two Governments agree that further meetings between their representatives should take place.

7. The Dominion Governments of India and Pakistan accept the above terms and express the hope that a friendly solution will be reached.

THE INDUS WATERS TREATY

Many difficulties soon arose. Irreconcilable positions were taken and threatening statements made by both sides. In June 1949, Pakistan laid claim to an equitable share of all the waters common to India and Pakistan. Pakistan suggested their approaching the International Court if negotiations did not lead to an early solution. Many messages were exchanged between the prime ministers but the stalemate continued.

David E. Lilienthal, the former chairman of the Tennessee Valley Authority and of the USA Atomic Energy Commission, in an article that appeared in the *Collier's* magazine in August 1951, wrote on Indo-Pakistan relations, Kashmir and the Indus waters.[2] He suggested a solution which, in

---

[2] Lilienthal had visited India and Pakistan earlier in February 1951. While India did not know at that time, it seems that he had prior discussions with Secretary of

his view, would assure Pakistan's then existing use while taking care of India's future development. He proposed 'that India and Pakistan work out a programme jointly to develop and jointly operate the Indus basin river system, upon which both nations were dependent for irrigation water. With new dams and irrigation canals, the Indus and its tributaries could be made to yield the additional water each country needed for increased food production'.[3] He suggested that the World Bank could be involved in its financing.

The World Bank showed interest and offered its good offices to both India and Pakistan in the matter. Eugene R. Black, the then president of the World Bank, wrote about it in September 1951 to the prime ministers of both India and Pakistan. Both countries generally accepted the offer.[4]

In March 1952, it was agreed that 'working parties' from India and Pakistan would prepare a comprehensive long-range plan for the entire Indus system of rivers. The aim was that it would secure the most effective utilization of water resources of the system and provide each side with irrigation uses substantially beyond what each side was enjoying then. Details of further consultations are not enumerated here but have been well-documented elsewhere.[5] No agreed plan emerged. However, it must be mentioned that the World Bank made its own proposals to both the countries in February 1954 and these were acceptable to India. These, *inter alia*, proposed that the waters of the Ravi, Beas, and Sutlej, should become available for the exclusive use of India.

Meanwhile, due to the uncertainties about the waters, a question arose on whether the work on the BNP should be suspended till a clear decision had been arrived at between India and Pakistan. India, however, decided to proceed with the construction even at the risk of losing a part of the investment, if warranted by the final settlement.

State Dean Acheson and President Truman, who encouraged him to make the visit. In fact, when Nehru visited the USA in October 1949, he met Lilienthal and extended an invitation to him to visit India.

[3] Quoted by N.D. Gulhati, in his book—*Indus Waters Treaty An Exercise in International Mediation*, Allied Publishers, New Delhi, 1973 from Lilienthal's article and *The Journals of David E.Lilienthal*; vol. III, Harper & Row, New York; 1966.

[4] There were a few demurrers. However, Prime Minister Nehru referred to Bhakra-Nangal and stated that 'this project will of course have to continue'. Pakistan was worried about its legal rights to continue its existing use of water.

[5] For instance, see Gulhati, N.D., 1973.

The Indus Waters Treaty that became effective from April 1960, resolved the water-related problems of the Indus basin.[6] Simply stated, after a transition period of ten years, this treaty awarded the waters of the eastern rivers, namely, the Ravi, the Beas, and the Sutlej, for the unrestricted use of India, except for some very limited use by Pakistan. Similarly, it allocated the waters of the western rivers namely, the Indus, the Jhelum, and the Chenab, for use by Pakistan and for certain uses by India. There was to be no quantitative limit on India's use of the western rivers for domestic and industrial purposes. Provisions for allowing their non-consumptive use in India were also made. A permanent Indus Commission was created for the implementation of the treaty and to serve as a regular channel of communication and contact between the two governments. Stringent and elaborate provisions for the resolution of differences and disputes were also made. The treaty was to continue in force until terminated by a duly ratified treaty for that purpose between the two governments.

The treaty stipulated that during a 'transition period', India should limit its withdrawals for agricultural use, limit abstractions for storages, and make deliveries to Pakistan from the eastern rivers in a specified manner. The limits of withdrawals from the Sutlej and Beas at Bhakra, Nangal, Ropar, Harike, and Ferozpur (including abstractions for storage by the Bhakra dam and for the ponds at Nangal and Harike), were specified by sub-periods of the year. The transition period was to end in 1970 (but was extendable up to 1973 at Pakistan's request). These imposed limitations up to March 1970.

The three eastern rivers, the waters of which were allocated for use by India, carry, on an average, about 40.5 BCM (32.8 MAF). Of this, the Sutlej accounts for 16.8 BCM (13.6 MAF). The BNP is designed to fully utilize the available waters of the Sutlej river allocated to India.

The role of the World Bank started as a 'good officer' but over the years its function actually went beyond 'good offices' or 'mediation' in the technical sense of those terms. It started playing a more active part (termed by some as 'continuing conciliation') in working out a solution.[7] It must be noted that it was because of this extended role that ultimately, the Bank made its own proposals to both the countries in February 1954. The Bank *inter alia*, proposed that the waters of the Ravi, Beas, and, Sutlej should become available for the exclusive use of India and that the waters of the Indus,

[6] For more details, see the text of *The Indus Waters Treaty-1960* along with the annexures.

[7] See Baxter, Richard R., 'The Indus River', in *The Law of International Drainage Basins*, A.H. Garretson, R.D. Hayton, and C.J. Olmstead (eds), 1967, p. 477 and Gulhati, N.D. 1973, p. 331.

Jhelum, and Chenab should be for Pakistan. A system of 'replacement works', a transitional period for this, as also a contribution by India towards the replacement works, were also envisaged. The treaty was largely based on these broad lines. An agreed draft emerged only by early September 1960. The agreement did not optimize the use of the waters of the Indus system, but resolved the irritants between the two nations at that time. This feeling was voiced over the years and hopefully the two nations will some day find ways and means of considering optimization of the system's potential in the future.

On 19 September 1960, the treaty was signed at Karachi by Field Marshal Mohammad Ayub Khan, President of Pakistan, and Jawaharlal Nehru, Prime Minister of India, and 'for the purposes specified' in some articles of the treaty, by William Iliff of the World Bank. It was ratified by both countries in December 1960 and became effective from 1 April 1960. The relevant details of the treaty are reproduced in Annexure-2.

DEVELOPMENT OF THE RAVI-BEAS WATERS

The waters of the Ravi and Beas rivers became available for use by India after 1970 under the Indus Waters Treaty. India took many advance steps in anticipation of this possibility. The states concerned arrived at an agreement on utilizing the surplus waters of the Ravi and the Beas in January 1955. The details of the Agreement are given in Annexure-3.

The mean supplies in the Ravi and Beas, over and above the actual pre-Partition uses (based on 1921–45 mean annual flow series), were calculated as under:

	Availability	
Beas	at Mandi plain	11.24 MAF
Ravi	at Madhopur	4.61 MAF
	Total	15.85 MAF

The allocation made of these waters under the 1955 Agreement was as under:

Punjab	5.90 MAF
PEPSU	1.30 MAF
Kashmir	0.65 MAF
Rajasthan	8.00 MAF
Total	15.85 MAF

In case of any variation in the total supplies, the shares were to change *pro rata*, except that the share of Kashmir would continue to remain at 0.65 MAF.

On the merger of the Patiala and East Punjab States Union (PEPSU) with Punjab in 1956, the share of the combined Punjab state stood at 7.20 MAF. Discussions on the splitting of the share allocated to this state, consequent to the reorganization of Punjab and the carving out of Haryana as a separate state, commenced in 1966.The Punjab Reorganization Act, 1966 contained special provisions in its sections 78, 79, and 80, for regulating the rights and liabilities of the successor states in relation to the Bhakra-Nangal and Beas projects. The BNP had been completed but the Beas project was still under execution. The Act cast the duty of completing it on behalf of the successor states upon the Union government with funds provided by the three states. The BNP management was entrusted to a new Bhakra Management Board.

Punjab and Haryana wanted their respective shares out of the 7.2 MAF, which was earlier allocated to united Punjab. Serious differences surfaced between these newly formed states regarding their respective shares in the Ravi and Beas river waters. Punjab claimed the entire portion for itself, as it contended that Haryana was no more a riparian state and no part of it was lying in the basin of these two rivers. Haryana contested this claim. The Government of India took the necessary steps in accordance with the Punjab Reorganization Act to determine the respective shares of these states. A notification was accordingly issued by the Government of India on 24 March 1976. As per this notification, Haryana was allocated 3.5 MAF and Punjab the remaining quantity not exceeding 3.5 MAF, after setting aside 0.20 MAF for Delhi's drinking water supply. This allocation was challenged in the Supreme Court of India. While the matter was pending before the Court, an agreement was entered into between Punjab, Haryana, and Rajasthan on 31 December 1981. The details of the Agreement are also given in Annexure-3. However, the issues were opened up once again due to subsequent political developments. Further negotiations were held and these led to the constitution of the Ravi-Beas Water Disputes Tribunal in 1986. The matter has not yet been finally resolved. However, these did not affect the BNP.

The integrated operation of the Bhakra-Nangal and the Beas Project (units I and II), as well as the Thein (Ranjit Sagar) dam, emerged as the bouquet of schemes to optimally utilize the Indus waters allocated to India under the Indus Treaty.

THE BEAS PROJECT

The Beas Project comprises two units, viz. Unit I—Beas-Sutlej Link, and Unit II—Beas dam at Pong. This project was implemented as a joint venture of the states of Punjab, Haryana, and Rajasthan, to extend and develop irrigation through the waters of the Beas river.

THE BEAS-SUTLEJ LINK (BSL) PROJECT

The general layout of the BSL Project is given in Figure 3.1.

The BSL involves a 76.2 m (250 ft) high and 255 m (836 ft) long earth-cum-rockfill dam at Pandoh situated 21 km (13 miles) upstream of Mandi town on the Beas river, to enable the diversion of 245.8 cumec (9000 cusec) of the Beas flows to the Sutlej. The link comprises the concrete-lined Pandoh-Buggi tunnel which is 13.1 km (8.1 miles) long at the start, thereafter the Sundarnagar hydel channel, which is 11.8 km (7.3 miles) long, a balancing reservoir of 3.7 million m³ (3000 acre ft) capacity, and the Sundarnagar–Sutlej tunnel, which is 12.35 km (7.7 miles) long at the end. The Dehar power station on the right bank of the Sutlej river is fed by the BSL at its tail. The powerhouse utilizes the head of 320 m (1050 ft) to generate hydropower at the Dehar power station with an installed capacity of 990 MW (6 units of 165 MW each). This flow further augments the generation of firm power at the Bhakra dam by 148 MW and is, thereafter, utilized for irrigation through the Bhakra canal system.

The total cost of Unit-I was Rs 449.17 crores. The cost of the Pandoh dam alone was Rs 45.43 crores.

PONG DAM

Under Unit-II, a dam was built across the Beas river at Pong, about 40 km (25 miles) from Mukerian in the Kangra district of Himachal Pradesh. It is an earth-core gravel shell dam, 132.66 m (435 ft) high and 1950 m (6400 ft) long. It impounds 8.57 BCM (6.95 MAC ft) of water, of which 7.29 BCM (5.91 MAC ft) is the live storage.

The project is primarily envisaged for meeting the irrigation water needs of Rajasthan, Punjab, and Haryana and is used for power generation as well.

The power plant has an installed capacity of 360 MW (six units of 60 MW) each. The annual irrigation is 1.6 million ha (4 million acres).

The total cost of Unit-II is Rs 325.88 crores, while the cost of the dam alone is Rs 38.10 crores.

MADHOPUR-BEAS LINK

It was proposed that the surplus waters of the Ravi, after meeting the needs of the UBDC, should be diverted to the Beas by constructing the Madhopur-Beas Link (originally with a capacity of 170 m³/sec, later raised to 227 m³/sec) to cater to the needs of the canals, dependent on the Beas and Sutlej flows. This link with a 283 m³/sec capacity was completed in 1955.

Figure 3.2 shows the rivers Sutlej, Beas, and Ravi and the inter-connected canals.

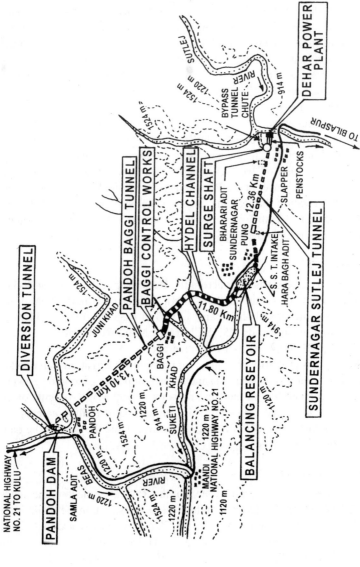

Figure 3.1: Beas-Sutlej Link (General Layout)

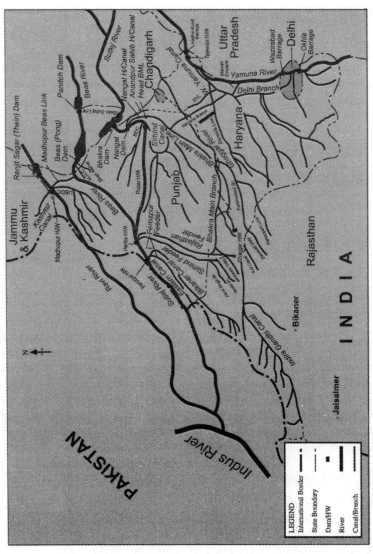

Figure 3.2: Sutlej, Beas, Ravi rivers and Connected Main Canals

Post-Independence Developments in Pakistan

New Dams

After the Indus Waters Treaty was signed, Pakistan undertook the construction of two storage reservoirs with the assistance of the World Bank, under the Indus Basin Development Fund. These were the Mangala dam and the Tarbela dam.

The Mangala dam is a 116 m high earth fill dam across the Jhelum river, 3058 m long with a controlled spillway. The live storage capacity is 6.58 BCM and the gross storage capacity 7.25 BCM. This work was completed in 1967.

The Tarbela dam is a 148 m high earth and rock dam, with a gated spillway, 2753 m long, built across the Indus river. It has a live storage of 11.46 BCM and a gross storage capacity of 13.68 BCM. This work was completed in 1976.

The main objectives of both these dams were the supplementation and regulation of irrigation water supplies and the generation of hydropower.

New Link Canals

After the Indus Treaty, in addition to the Mangala and Tarbela dams, Pakistan also undertook 'replacement works' in order to convey the waters of the Indus, Jhelum, and Chenab rivers to the Sutlej—at Suleimanki and below Islam—for replacing the waters of the Ravi, Beas, and Sutlej, which were 'reallocated under the Indus Waters Treaty'. Pakistan undertook the construction of some new barrages, as well as the remodelling of existing ones, so that the requisite flows to the new link canals could be diverted.

The link channels were as under:

Link Channel	*Linked systems/rivers*
1. Chashma barrage—Jhelum	Indus and Jhelum
2. Taunsa barrage—Panjnad	Indus and Panjnad
3. Rasul barrage—Qadiabad	Jhelum and Chenab
4. Balloki—Suleimanki	Ravi and Sutlej
5. Trimmu barrage—Sidhnai	Chenab and Ravi
6. Sidhnai barrage—Mailsi	Ravi and Sutlej
7. Mailsi siphon-barrage—Bahawal	Link with old canal

Figure 3.3 is a diagrammatic representation of these links.

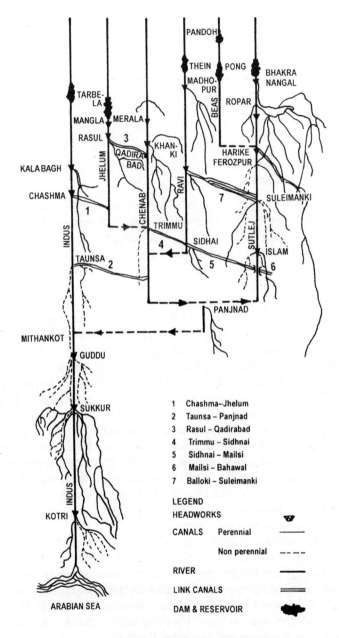

Figure 3.3: Projects Developed Under Indus Water Treaty

4

THE PRE-PROJECT SCENE

Post-Partition India

The province of Punjab was divided at the time of Independence into the new provinces of East Punjab and West Punjab. Seventeen districts of the undivided province were treated as constituting the new province of West Punjab, which became part of Pakistan. The remaining twelve districts were treated as forming East Punjab, which remained with India. The Punjab Boundary Commission Award given by Sir Cyril Radcliffe, split some of the districts, and made further modifications in the boundary between India and Pakistan. Radcliffe recognized that the task of delimiting the boundary in Punjab was indeed difficult. The finally drawn boundary cut across communities, rivers, and canals and left everyone in the region unhappy.

After Partition, while nearly half of the cultivable land in the Indus basin remained with India, only 20 per cent of the irrigated area was given to it. This left several million hectares of arid but fertile land in the Indian part of the Indus basin literally high and dry. Moreover, the allocation of a major part of the highly developed canal irrigated areas to West Punjab worsened the food problem for India. The sudden, mass arrival from West Punjab into India of millions of non-Muslim agriculturists as refugees aggravated the problems. The Government of India was compelled to act very quickly to ameliorate the situation. The early finalization and execution of the BNP was one of the important actions that independent India took to deal with the problems suddenly confronting her.

PEPSU was merged with East Punjab with effect from 1 November 1956 and the reorganized state was named Punjab. In November 1966, the state was bifurcated into Punjab and Haryana, and simultaneously the hill districts of Simla and Kangra were merged with Himachal Pradesh. However, throughout the period of the planning and execution of the BNP, it was the combined Punjab state (including Haryana) that was mainly concerned with it, apart from Rajasthan and Himachal Pradesh.

Between the Sutlej and Yamuna rivers lay the vast, alluvial, gently sloping plains, which formed the watershed between the Indus and the Ganga. The BNP aimed at bringing these areas under assured canal irrigation to protect the region from recurring droughts and famines.

Any study of the social and environmental impacts of a project undertaken in the early years of independent India is handicapped by the lack of requisite baseline data. This is understandable if we recall that till 1978 there was no formal requirement to conduct an environmental impact assessment (EIA) or to assess the social impact of a project. The Ministry of Environment and Forests issued guidelines for EIAs of river valley projects in 1985. Even these dealt mainly with upstream issues, but included questions relating to rehabilitation. This became a statutory requirement only in 1994.

Consequently, all such data which are now considered, with the benefit of hindsight, as relevant baseline information for studies of social and environmental impacts, were not often collected. Further, all collected data are not easily accessible in respect of projects completed many decades ago. Bhakra-Nangal is no exception to this. Nevertheless, an attempt has been made to put together the picture—in the years just prior to the start of the project—of the social, cultural, environmental, and economic scene that existed both in the catchment and reservoir zone upstream, as also the benefitted region downstream.

FAMINES

South-west Punjab and the adjoining areas of Rajasthan had suffered from many devastating famines as a result of severe drought till India's independence. Records speak of Emperor Akbar ordaining in 1568 that 'this jungle (Hisar district in Punjab) in which subsistence is obtained with thirst, be converted into a place of comfort'. Something was perhaps done but obviously that did not stop famines from recurring. The arrival of the British on the scene and their rule did not make much difference to this part of India, at least with regard to droughts and famines. The Government of India appointed three successive Famine Commissions in the last three decades of the nineteenth century to report on the dimensions of the droughts, their causes and on the relief to the affected people. Detailed codes for providing relief were drawn up. However, the Hisar district featured repeatedly in famine distress records. The *District Gazetteer* of Hisar states that the first such case for which authentic accounts existed relates to the year 1783. Many thousands died as a result. Similar accounts are narrated about the famines of 1860–1, 1868–9, 1876–8, 1899–1900, 1918, 1929–30, 1932–3 and 1938–40. Likewise, accounts are available in the *District Gazetteers* relating to the other districts of Punjab (for example, Ferozpur, Amritsar) and Rajputana (for example, Sriganganagar).

The Report of the Famine Commission (1860–1) discussed the famine tract south of the Sutlej (particularly Ambala, Patiala, and Bhatinda) as follows:

It is black, wretched, and without water, a mere sterile land. The wells are so deep that artificial irrigation from them is impossible; the water is so brackish and impure which none save natives of the tract can drink it with impunity; rains are scanty and precarious; vegetation is represented by a few stunted thorn bushes, or a temporary crop of grass, scattered over the great parched plains. Under circumstances so ungenial, the population is necessarily scanty and lawless...

The Famine Commission Reports trace the history of past famines from 1769. Since the 1870s, with the standardization of rain gauge networks and the availability of the India Meteorological Department (IMD) records, the documentation of droughts was placed on a sound footing. Unfortunately, it appears that a fatalistic view was taken by the administration that droughts and famines would be unavoidable, and all that could be done was to save human lives to the extent possible. It is, however, noted that even then, many voices were raised in favour of drought-proofing these regions. Some follow-up actions in providing canal irrigation in Punjab and Bikaner followed but not in the region that we are concerned with here. It was almost at the dawn of Independence that this long overdue action was initiated. It is indeed ironical that this part of India had to suffer for such a long period due to the loss of agricultural production and consequent famines.

DROUGHT-PRONE AREA

Drought is essentially a meteorological phenomenon. The vulnerability of any area to drought depends on the extent to which physical and climatic conditions play an adverse role in creating an unstable agriculture. The difference between chronic drought areas and drought areas is one of frequency. The Irrigation Commission defined drought-affected areas as those where the annual rainfall was less than a 100 cm and even 75 per cent of this rainfall was not received in 20 per cent or more of the years. Over large areas of Punjab (including Haryana) as also Rajasthan, the average annual rainfall is below 60 cm and the western parts of Rajasthan and adjoining areas in Punjab receive hardly 10 to 30 cm rain. The Irrigation Commission pointed out that irrigation, to the extent it could be provided, would give protection to many of the drought-prone areas. After adequate irrigation coverage is extended to these areas, they no longer need to be classified as 'drought affected'.

DRINKING WATER

Water sustains life and hence, is a basic need and right. The fundamental right to life is held to include the right to drinking water. From ancient times, the provision of water for domestic use has been considered to be an essential duty of the ruler. Unfortunately, large parts of the project region completely lacked this basic amenity till the BNP was taken up and completed. In particular, the areas in Rajasthan and the adjoining parts of Haryana and Punjab suffered acutely due to this problem. It is a typical sight in this area to watch women and children walking miles to fetch potable water for domestic use, deftly balancing a number of pots on their heads and in their hands. Similarly, at many places the menfolk go for miles in camel-drawn vehicles to fetch water from far-off sources. The Irrigation Commission pointed out that in some places in Rajasthan, 'even drinking water is not available and we are of the opinion that drinking water should be made available without considerations of cost'.[1] The present National Water Policy of India, too, stipulates that 'adequate safe drinking water facilities should be provided to the entire population both in urban and rural areas'. It further directs that 'drinking water needs of humans beings and animals should be the first charge on any available water'.[2]

PHYSIOGRAPHY

The project area lies in the Sutlej basin, forming part of the Indus system. The Great Himalaya, the Pir Panjal, and the Dhauladhar and Dhartidhar ranges of the Siwaliks enclose the watershed of the Sutlej. The river rises from the Great Himalaya in the Tibet Autonomous Region of China. It enters India in Himachal Pradesh, where it runs through Kinnaur, Shimla, Solan, Bilaspur, and Una districts. It then enters Punjab.

Punjab has two major physiographical divisions, namely the sub-montane tract and the alluvial plains. The upper parts of the districts of Gurdaspur, Hoshiarpur and Rupar lie in the sub-montane tract. The plains are part of the Indo-Gangetic plains.

Haryana is bounded by the Siwaliks in the north and the Yamuna in the east. The Aravali range runs along the south, while the desert tracts of Bikaner lie in the south-west.

[1] Government of India, Irrigation Commission Report, vol. II , Chapter XVII, p. 349, 1972.
[2] Government of India, Ministry of Water Resources, *National Water Policy*, para 8, 2002.

The Rajasthan plains west of the Aravali form a sandy semi-arid region. Isolated patches of scrub jungle break the monotony of the desert.

CLIMATE

The climate of the region varies from perpetually snow-capped peaks in parts of Himachal Pradesh to the tropical heat of the sub-montane and plain tracts of Punjab, (including Haryana), and Rajasthan. During summer, the temperatures soar and humidity drops. Day temperatures increase considerably, in association with heat waves and dust storms.

The climate in Haryana over most of the year is of a pronounced continental character, that is, very hot in summer and very cold in winter. The temperature in May–June rises to 40°C but drops to 2 to 3°C in January. During winter, the region remains under the influence of cool winds with occasional rains. Great heat, hot winds, and dust storms, particularly in water-scarce Hisar and Mahendragarh, characterize the summer.

Climatically, Punjab can be divided into three major seasons: (1) Summer (April to June); (ii) Monsoon (July to September); and (iii) Winter (October to March).

In the hot weather, the temperature rises to 43°C, associated with heat waves and occasional dust storms. During the cold weather, it goes down to 4°C and at times even below. Occasionally, there are winter rains. Altitude and rainfall govern the climate in Himachal Pradesh. Snow-covered mountain slopes are rugged and inhospitable. Lower down, there is snow in winter and rain in the monsoon.

In the paragraphs that follow, details regarding the rainfall, temperature, evaporation, wind speed, and relative humidity in the states of Punjab and Haryana are given. This is followed by corresponding details with respect to the Ganganagar area of Rajasthan. Together, these will represent the conditions in the project region.

RAINFALL

The annual rainfall in the Punjab plains varies between 40 and 80 cm, while it is higher in the sub-montane districts. The rainfall in Haryana is scanty and erratic. In the southern part, it is as low as 25 to 40 cm. Rajasthan is the driest state of India. The annual rainfall is as low as 13 cm in the Thar desert. The districts of Bikaner, Ganganagar, Barmer, Jodhpur, etc., fall in the arid and semi-arid zones. The plains districts of Punjab and Haryana and the Ganganagar district of northern Rajasthan were to be provided with assured irrigation, based on the storage at Bhakra. The rainfall pattern in the major districts in this region is indicated in Table 4.1.

Table 4.1: Rainfall Averages in select districts of Punjab and Haryana

(all figures in cm)

District	Average rain for 5 years (1955–59)				Average rain for 5 years (1996–2000)			
	January–May	June–October	November–December	Total	January–May	June–October	November–December	Total
Hisar	3.5	37.5	1.5	42.5	3.9	24.9	0.1	28.9
Sirsa	Part of Hisar dist. Till 1975				3.9	22.6	0.4	26.9
Fatehabad	Part of Hisar dist. Till 1997				5.8	31.0	2.1	38.9
Karnal	8.1	67.5	2.0	77.6	8.8	50.7	0.5	60.0
Jalandhar	11.1	73.8	2.9	87.8	13.9	53.9	1.3	69.1
Ludhiana	9.8	75.7	2.2	87.7	11.5	49.9	1.9	63.3
Ferozpur	6.1	35.9	1.9	43.9	6.5	22.6	0.1	29.2
Bhatinda	14.9	41.2	1.3	57.4	3.3	20.5	0.2	24.0

Source: 1. *Statistical Abstracts of Punjab and Haryana of various years.*
2. *Water and Power Consultancy Services (India) Ltd. (WAPCOS) records.*

TEMPERATURE

The temperatures recorded during the different months of 1999 with respect to representative stations in the command area are summarized and presented in Table 4.2.

EVAPORATION

Evaporation and transpiration closely follow the march of seasons. When the temperature is low and the wind speed slight, evaporation is low. As the day temperature rises, the cloud cover is low and the wind velocity increases, evaporation also increases. The peak rate of evaporation is in the summer months of May–June. With the advent of the monsoon, there is a marked fall in evaporation rates. The mean annual evaporation over the Beas-Sutlej catchment is of the order of 264 cm.

Table 4.3 gives the monthly break-up of the mean values of evaporation. The highest evaporation occurs in the month of May and the lowest in December.

Table 4.2: Temperature at select locations in Punjab and Haryana (1999)
(all figures in °C)

	Ludhiana			Patiala			Hisar		
	H	L	M	H	L	M	H	L	M
January	23.1	3.2	11.7	24.0	5.2	12.6	21.8	6.0	12.7
February	26.0	2.8	15.7	25.8	3.4	15.8	27.3	4.5	17.2
March	36.3	7.7	20.6	37.0	8.6	21.1	39.5	10.7	23.5
April	43.4	12.4	28.6	43.4	11.7	28.7	45.4	15.9	31.6
May	44.4	19.6	31.6	43.8	19.3	31.9	47.8	21.8	34.4
June	41.1	20.0	30.8	41.4	19.1	31.5	45.3	20.3	34.1
July	36.9	20.6	29.6	37.7	21.9	30.4	44.9	25.1	34.0
August	35.0	20.9	28.9	35.4	23.1	29.7	40.1	25.3	32.7
September	36.0	21.6	28.6	35.8	21.5	28.9	39.3	24.3	31.9
October	33.8	14.0	24.5	34.8	14.5	25.1	36.1	16.4	27.1
November	33.7	7.0	19.4	32.6	7.5	19.8	34.0	10.0	16.4
December	26.2	0.1	13.7	26.8	3.2	14.4	27.6	4.1	15.9

Source: Regional Meteorological Centre, New Delhi, Statistical Abstract of Punjab (2000) (p. 146) and Statistical Abstract of Haryana (2001) (pp. 98–101).
Notes: H = Highest at any point of time during the month.
L = Lowest at any point of time during the month.
M =1/2 (mean daily maximum plus mean daily minimum during the month).

Table 4.3: Mean monthly evaporation in the Beas-Sutlej catchment

(all figures in cm)

Month	Evaporation	Month	Evaporation
January	9.6	July	22.6
February	14.2	August	18.3
March	23.5	September	18.9
April	30.9	October	18.6
May	43.1	November	13.5
June	41.4	December	9.2

Source: WAPCOS.

Note: The years and the stations considered are not given by WAPCOS.

SURFACE WIND SPEED

The figures of average wind speed in the various months of 1999 at select locations in Punjab and Haryana are given in Table 4.4

Table 4.4: Average wind speed (1999)

(all figures in km/hr)

Month	Location			Month	Location		
	Patiala	Hisar	Ambala		Patiala	Hisar	Ambala
January	3.5	2.7	3.2	July	3.5	4.0	3.9
February	3.8	3.3	3.7	August	2.7	3.9	3.5
March	4.3	4.3	5.2	September	1.6	3.2	2.7
April	3.2	3.6	4.4	October	1.5	1.8	1.9
May	5.1	4.6	4.9	November	2.4	1.8	2.8
June	5.1	4.9	4.9	December	1.5	1.3	1.6

Source: Regional Meteorological Centre, New Delhi, Statistical Abstracts of Punjab (2000) and Haryana (2001).

RELATIVE HUMIDITY

The mean relative humidity in different months of 1999 at select areas in Punjab and Haryana are given in Table 4.5.

Ganganagar is immediately to the west of the Punjab and Haryana region discussed above. It adjoins the arid desert and suffers periodically from drought and consequent distress. Even drinking water is not available. The details of rainfall, temperature, relative humidity, and wind speed for Ganganagar are given in Table 4.6.

Table 4.5: Mean Relative Humidity (1999)

(all figures in %)

	Ludhiana	Patiala	Hisar	Ambala
January	87	87	96	88
February	73	77	89	90
March	53	58	68	64
April	23	26	34	43
May	35	43	49	51
June	54	58	60	67
July	76	75	69	84
August	76	78	70	84
September	76	78	74	85
October	63	80	73	81
November	59	65	71	78
December	79	84	92	91

Source: Regional Meteorological Centre, New Delhi, Statistical Abstracts of Punjab (2000) and Haryana (2001).

Table 4.6: Climate Statistics for Ganganagar

Month	Mean rainfall (cm)	Air temperature in the month		Mean relative humidity (%)	Mean wind speed (km/hr)
		Highest (°C)	Lowest (°C)		
January	1.4	25.2	0.6	48	4.0
February	0.9	27.6	2.1	55	5.0
March	1.4	35.8	6.3	47	6.4
April	0.5	42.1	12.5	31	6.8
May	1.1	46.1	18.2	25	8.0
June	3.6	46.2	22.3	36	10.7
July	6.9	43.5	24.0	55	13.3
August	7.7	40.6	23.8	64	8.0
September	4.0	40.2	19.9	54	6.2
October	0.3	38.2	11.7	43	4.7
November	0.2	33.6	5.5	47	3.3
December	0.7	27.8	2.0	59	3.4

Source: Government of Rajasthan, Irrigation Department.
Note: The year to which the figures relate is not given in the source document.

GEOLOGY OF THE DAM REGION

The catchment and the reservoir are in the Himalayas, with the Bhakra and Nangal sites representing almost the last points in the hills before the river enters the plains. The dam site is located in the Nahan series of sandstone, alternating with clay-stone strata. The Bhakra gorge is cut into members of the Siwalik formation which have basal beds preponderantly of sandstone but with important clay-stone and/or silt-stone layers intervening with the sandstone horizons.[3] Stratification is almost at a right angle to the gorge. The dam has been located so that it rests upon the largest member of the sandstone series. Upstream and downstream of the dam, there are wide clay-stone bands within narrow boundaries.

SOILS

The hill districts of old Himachal Pradesh, namely, Kinnaur, Mahasu, Mandi, Chamba, and Bilaspur, constitute five soil zones. In Mandi and a greater portion of Bilaspur, low hill type soils suitable for growing wheat, maize, ginger, paddy, and citrus fruit are found. The middle hill type soils found in Rampur, Kasumpati, and Solan tehsils of Mahasu are suitable for growing potato, wheat, and maize. The hill type soils are good for seed potatoes and temperate fruits. Mountain soils unsuitable for agriculture are found in parts of Mandi. Tibet and the Pangi region of Himachal Pradesh get heavy snow. They have dry hill soil over sandstone, suitable for fruit orchards.

The sub-montane region of Punjab has forest and hill soils ranging from lightly acidic to highly acidic in reaction, which are in different stages of podzolization. Though rich in humus, they contain very little soluble salts and are somewhat deficient in lime and phosphoric acid. The districts of Ferozpur and Bhatinda have desert soils, which not only lack moisture but are also deficient in organic matter, nitrogen, and phosphorus. The predominant soil groups in the state of Punjab belong to the Indo-Gangetic alluvium class. The soil crust contains sodium salts. The soils are deficient in nitrogen and organic matter.

In Haryana, too, different types of soils are found. Desert soils are found in parts of Hisar and Mahendragarh where the annual rainfall is less than 30 cm. These soils are deficient in nitrogen, phosphorus, and potash. Sierozan soils are found in parts of Hisar and Mahendragarh where the rainfall is between 30 and 50 cm per year. Salinity and alkalinity could become

[3] Handa, C.L. and Chadha, O.P. , 'Bhakra Dam Salient Design Features', *Indian Journal of River Valley Development*, November 1953, p. 25.

problems in such areas. These soils too are deficient in nitrogen, phosphorus, and potash. Erosion by wind is common.

Arid brown soils are found in Karnal, Rohtak, and Jind districts, where the annual rainfall is between 50 and 65 cm. Problems of wind and water erosion are common. These soils are also deficient in nitrogen but contain phosphorus and potash.

Tropical arid brown soils are found in parts of Karnal and Ambala districts, where the rainfall is 75–90 cm. They are deficient in nitrogen, phosphorus, and potash.

LAND UTILIZATION

IN THE COMMAND AREA

The total extent of cultivable area in post-Partition Punjab (excluding Simla and Kangra districts) in the early 1950s was around 8.5 million ha. The net area sown was around 6.6 million ha. The net irrigated area was about 43 per cent. The annual figures for the five years immediately prior to the commencement of irrigation from the Nangal barrage are detailed in Table 4.7.

Table 4.7: Development of irrigation in post-Partition Punjab (1950–1 to 1954–5)
(all figures in 1000 ha)

1	2	3	4	5	6	7	8	9
Year	Net area sown	Col. 2 as % of total area	Total cropped area	Col. 4 as % of col. 2	Net irrigated area	Col. 6 as % of col. 2	Gross irrigated area	Col. 8 as % of col. 2
1950–51	6803	72	8059	118	2589	38	3006	44
1951–52	6421	68	7480	116	2636	41	3175	49
1952–53	6489	69	7638	118	2881	44	3258	50
1953–54	6386	68	8164	128	3025	44	3360	53
1954–55	7050	75	8966	127	3287	47	3622	51
Average	6720	71	8061	120	2883	43	3284	49

Source: Statistical Abstract of Punjab.

There were problems with the type and quality of irrigation provided. The canal-irrigated area accounted for just one-fourth of the net area sown. Tube-wells were yet to appear in any significant manner. There was neither electricity nor diesel for running the pumps. Well irrigation, which accounted for a third of the irrigated area, was practised with the aid of bullock-driven Persian wheels or water buckets lifted with camel power. The intensity of irrigation and water allowance were very low. In the absence of storage, the canal systems were dependent upon the flow of the river. In most years, during the non-monsoon months, the deliveries at the canal heads dwindled. In drought years, it became impossible to deliver 'authorized' full supply discharges, even in the rainy season. The deliveries of water were unsteady and unreliable. Thus, even with the 'available irrigation', agriculture and food production remained a gamble with the rain.

CROP YIELDS

Since the cropping pattern was predominantly governed by rain, the yields were low. Moreover, the use of high-yield variety (HYV) seeds or fertilizers had not developed. The yields of principal crops during the five years between 1950–1 and 1954–5 are shown in Table 4.8.

Table 4.8: Yields of crops in Punjab 1950–1 to 1954–5

(yield in kg per ha)

Year	Rice	Wheat	Maize	Foodgrains	Oilseeds	Cotton
1950–1	729	860	580	590	400	1140
1951–2	805	960	920	630	340	1220
1952–3	1001	1100	1370	810	420	1360
1953–4	995	1060	1343	850	450	1550
1954–5	955	1100	1160	710	450	1290
Average	897	1010	1070	730	400	1310

Source: Statistical Abstract of Punjab.
Note: Cotton yield is in the form of bales per ha.

The people of south-west Punjab resorted to jawar, bajra etc., for food grains, as the level of wheat production was low. While undivided Punjab ranked as the granary for India, after Partition, Punjab became deficient in food grains.

FLOODS

In addition to suffering from famines during the drought years, Punjab was devastated by flood, too, in wet years. There are a number of small and big hill torrents that spill into the plains during flood. Soon after the rivers enter the plains, they spread out and start meandering. The sudden change in the gradient aggravates the problems. The shifting of river courses added to the miseries of the people. Very often, the swollen rivers inundated fertile cultivated lands which were deposited with sand. Communication lines were disrupted and villages were eroded away.

The national flood policy statement of 1954 viewed that 'construction of storage reservoirs is among the most effective measures for the control of floods'. The High Level Committee on Floods (1957), *inter alia*, recommended that flood control schemes should fit in with plans for other water resource development such as irrigation, power, and domestic water supply, to the extent possible.

IN THE SUBMERGENCE AREA

The submergence area acquired by the project was 17,984 ha. In this area, 'forest land' accounted for 5747 ha. Moreover, though classified as 'forest', much of it was only degraded forest. *Khair, shisham*, mango, *slambra*, and *kitcham* were reported as the trees prevalent in this area. Some naturally occurring species of medicinal plants reported are *amaltas, amla, neem, bahara*, and *harar*.

About 10,000 ha of land in the submergence area were reportedly under cultivation without any irrigation, and the yields were low.

PLANT SPECIES IN THE CATCHMENT

The plant species depends upon the altitude, climate and soil of specific regions. An overall picture is reported in Table 4.9.

Table 4.9: Plant species in the catchment

S. no	Elevation	Species
1	Below 900 m	Mainly scrub forests. Important species are *khair, shisham*, bamboo, *acacia, grewia*, mulberry
2	900–1525 m	Mainly chiripine
3	1525–2440 m	Mainly *deodhar, kail* and *ban*
4	2440–3300 m	Mainly spruce and fir forests
5	Above 3300 m	No vegetation except herbs and grasses

Source: BBMB communication.

WATER QUALITY

There are no details given in the project report about the quality of the waters of the Sutlej river. However, it is generally known that the water was potable and good in quality. The results of the water quality analyses of the post-project period are available and the quality of water can now be termed as good. The details of some recent water quality analyses are given in Table 4.10

FISHERIES

No details about the fish population in the pre-project days have been given in the project reports. It has been presumed that not much fishing existed in those days. However, detailed information on the present position is available. Some fifty-one species of fish, belonging to eight families, have been recorded in the Gobindsagar reservoir by the Central Inland Fisheries Research Institute. These families are: Cyprindae, Cobitidae, Bagridae, Schilbeidae, Sisoridae, Belonidae, Ophiocephalidae, and Mastacembelidae.

The present average catch per day is three tonnes or annually 1100 tonnes. About 1400 fishermen are engaged full-time in fishing activities in the reservoir. The corresponding data of pre-project years is unavailable.

COMMUNICATIONS

Hilly regions in India had always been remote and inaccessible. Roads and railways were non-existent. The BNP region was no exception. The Sutlej catchment area upstream of the Sirhind canal headworks in Ropar were largely inaccessible. In the absence of communications, these remote areas were deprived of development and even the minimum standards of the quality of life for the people of the region.

H.W. Nicholson, who investigated the site during 1915–18, has recorded that he used to proceed to Bhakra from Ropar. He would hire a bullock cart and sleep in it, while the cart moved overnight from Ropar to Nangal. Those who made the journey on the right bank of the river took even longer. B.R. Palta, who made investigations in 1939–42, has recorded that it took one day to go from Lahore to Una via Hoshiarpur. The Soan river at the outskirts of Una was unbridged and one had to wade through waist-deep water to reach Una. At Una, a horse was hired to go to the Nangal ferry. From Nangal, he used to walk over the hills by a footpath to Bhakra. During the time of the investigations in 1946, there was hardly any road from Ropar to Nangal. A Ropar-based syndicate operated a bus through muddy, sandy, deeply rutted *katcha* roads, with often shifting alignments. During the monsoon season, crossing the hill torrents was treacherous and many lives were lost during these attempts.

Bhakra remained an inaccessible spot even in 1947. It took two days for the project investigating team and inspecting officers to reach the Bhakra dam site even then. The surveys upstream in the reservoir area meant going

Table 4.10: Test results of water samples collected from the Bhakra dam region

S. no.	Parametres	Units	Specified limits as per BIS	X-5	X-27	X-51	X-69	X-133	D/s Bhakra PP-II
1	PH Value	Mg/l	6.6–8.6	8.3	8.6	8.6	8.6	8.6	8.1
2	D.O.	Mg/l		9.2	9.8	7.6	7.0	9.4	9.2
3	B.O.D	Mg/l	2.0	0.7	2.3	1.8	0.7	1.8	0.7
4	C.O.D	Mg/l		10.0	26.6	20.4	14.8	18.8	13.2
5	Dissolved Solids	Mg/l	500.0	141.0	141.0	141.0	137.0	171.0	169.0
6	Chlorides	Mg/l	260.0	29.0	30.0	30.0	32.0	30.0	29.0
7	Sulphates	Mg/l	400.0	6.0	4.0	4.0	4.0	4.0	6.0
8	Hardness Total	Mg/l	300.0	130.0	122.0	112.0	124.0	138.0	146.0
9	Nitrate	Mg/l	20.0	ND	ND	ND	ND	ND	ND
10	Suspended solids	Mg/l		ND	ND	ND	ND	ND	ND
11	Ammonical Nitrogen	Mg/l	Nil	ND	ND	ND	ND	ND	ND
12	Sodium	Mg/l		2.5	2.6	2.6	2.4	6.1	4.0
13	Dissolved Phosphates	Mg/l		0.800	0.700	0.700	0.300	0.700	0.500
14	Conductivity	Micro-sec/cm	300.0	166.5	167.0	157.4	162.6	190.4	196.1
15	Alkalinity	Mg/l		79.5	91.5	68.0	64.0	92.6	74.5

Source: BBMB.

Note: Tests were conducted by the Haryana State Pollution Control Board, except for D.O. which is taken from tests carried out at Roorkee. The dates of sampling and testing were not indicated in the source document.

N.D. = No data.

into even more inaccessible territory. Jagman Singh reported that even around 1949, the officer in charge of the right diversion tunnel 'used to walk from Nangal to his place of work, right tunnel, a distance of about 10 km through treacherous and tortuous foot paths. He thus walked more than 20 km everyday'.[4]

A brief account of how the poor communication facilities existing in those early days were improved upon by the project may be recounted below.

RAILWAYS

Prior to the decision on taking up the Nangal barrage, the nearest rail terminus was at Ropar, 69 km (43 miles) away from the project site. The extension of the railway line along the left bank of the river till Nangal and beyond was decided upon to facilitate the movement of workers and materials. The construction of the railway line was entrusted to the Northern Railway. The route was through very difficult terrain and involved crossing over 60 torrents by bridges. The Ropar-Nangal rail link was opened to traffic in October 1948.

The railways did not consider the section beyond the Nangal barrage to Bhakra as commercially viable. Hence, the Nangal-Bhakra link was built by the project. It was 27 km (17 miles) long, including the sidings.

A rail-cum-road bridge was built across the Sutlej at Olinda in 1953 to provide a link to the other bank. Till then, the only communication to the right side of the river from the left bank approach road was by two suspension bridges built by the project. The railway line on the right side was laid in 1953.

ROADS

Construction of the highway from Ropar to Nangal was also undertaken around the same time. The 58 km (36 miles) long highway from Ropar to Nangal was constructed by the PWD Punjab and was completed by 1947. The road from Nangal to Bhakra, about 11 km (7 miles) long, was constructed on the left side of the Sutlej river by the Bhakra project administration. This was a 7.6 m (25 ft.) wide hill road.

POSTS AND TELEGRAPH

There were no regular postal, telegraph or telephone facilities upstream of Ropar town till work on the BNP started. A small sub-post office was

[4] Singh, Jagman, *My Trysts with the Projects-Bhakra and Beas*, Uppal Publishing House, New Delhi, 1998, pp. 91–2.

opened in the Nangal township with telephone and telegraph facilities in 1947. Later, two more sub-post offices were opened at the Nangal barrage and labour colony. It was only later that an efficient telegraph system through a PBX manual exchange started working.

MEDICAL FACILITIES AND HEALTH

A health impact assessment is as much a part of a project assessment as the economic, social and environmental aspects. The health status of the project region, as well as the upstream and downstream areas in the pre-project days has, therefore, to be assessed in detail so as to monitor subsequent changes. The project report does not give details in this regard. Pre-existing nutritional and health conditions had to be surmised. Water-borne diseases like cholera, diarrhoea, and dysentery were known to be prevalent in all the areas which had poor quality drinking water.

As in other hilly and remote locations in India, there were hardly any modern hospitals or health facilities in the project region and upstream areas. In the absence of road or rail facilities, if someone fell seriously ill, it was a long, hazardous procedure involving many days, to take the patient to the nearest hospital with some modern health-care facilities. Even during the days of investigations and the initial days of the project construction, these facilities were unavailable to the project staff or other people of the township till 1951. In general, it can be stated that access to drinking water is an important step towards better health.

MALARIA AND MOSQUITOES

It has often been stated that the development of irrigation has led to the proliferation of mosquitoes and malaria. In order to assess whether a project made any difference to this picture, data on the baseline situation that existed in the pre-project days, as well as on the post-project scene, are needed. Unfortunately, such pre-project data is not available in many cases, including with respect to the BNP. Even if these were available, a practical difficulty arises in precisely indicating how much of the change, if any, was attributable to the project. The history of the control of mosquitoes and malaria is, itself, an interesting story.[5]

Malaria has plagued mankind for countless generations. Till the early twentieth century, malaria was a dreaded disease in India. Vast tracts of land

[5] For instance see Tren, Richard, and Bate, Roger, *When Politics kills- Malaria and the DDT Story,* Occasional Paper 7, Liberty Institute, New Delhi, 2000.

were uninhabitable because of it. The link between human beings, malaria, and the anopheles mosquito was discovered in 1898 and, hence, in addition to treating the disease, a programme for the control of mosquitoes was launched. The next major advance in vector control came in the form of DDT spraying in the late 1940s and 1950s. With widespread spraying, the malaria eradication programme nearly succeeded. Against 75 million malaria cases reported in 1951, the figure came down to a mere 50,000 in 1961. Against 800,000 deaths in 1951, there were only a few hundreds in 1961.[6]

In the 1960s, a scare was created that the use of DDT was having a devastating effect on bird life. By 1972, the US environment protection agency banned the use of DDT. Most Western countries followed suit and they enforced a near ban on its use because of the environmental scare. Mosquitoes, too, hit back and malaria returned.

GREENHOUSE GASES (GHGs)

The emission of GHG from reservoirs, due to rotting vegetation and carbon inflows from the catchment, is a recently identified eco-system impact. Such emissions can change significantly over time. It is known that GHG emissions from deep reservoirs are much less than those from shallow reservoirs. Data on pre-dam GHG emissions are not available for the BNP. Such figures were, in fact, not available with respect to any of the dams studied by the WCD.

---

[6] Harrison, G., *Mosquitoes, Malaria and Man: A History of the Hostilities since 1880*, John Murray, London, 1978, p. 247.

5

BHAKRA-NANGAL PROJECT

'The Biggest Places of Worship'

Prime Minister Jawaharlal Nehru made an inspired speech at Nangal on 8 July 1954, at the opening of the Nangal canal. Just before coming to Nangal, he was in Bhakra, where the Bhakra dam was still being built. Standing on the banks of the Sutlej, he saw the people working at various spots. For Nehru, nothing was more encouraging than the sight of people trying to capture India's dreams and giving them real shape. He felt that Bhakra-Nangal was such a place, a landmark, because it had become the symbol of a nation's will to march forward with strength, determination, and courage. He told his audience that Bhakra-Nangal is 'not only for our own times but for coming generations and future times'.

Excerpts from this speech are often quoted. These have also been used pejoratively by some critics who state that Nehru called, with religious awe and reverence, large dams as 'Temples of modern India'.[1] It is, thus, useful to first quote correctly what he had stated.[2] This is the relevant extract:

As I walked round the site, I thought that these days the biggest temple and mosque and *gurdwara* is the place where man works for the good of mankind. Which place can be greater than this, the Bhakra-Nangal, where thousands and lakhs of men have worked, have shed their blood and sweat and laid down their lives as well? Where can be a greater and holier place than this, which we can regard as higher?

[1] For instance, Arundhati Roy, in her essay against large dams entitled, 'The Greater Common Good' said, 'In the fifty years since independence, after Nehru's famous 'Dams are the Temples of Modern India' speech, (one he grew to regret in his own lifetime), his foot soldiers threw themselves into the business of building dams with unnatural fervour'. McCully mentions 'Religious Reverence'.

[2] Speeches of Jawaharlal Nehru—See CWC publication no. 49/89, Jawaharlal Nehru and Water Resource Development (Some of his speeches: a compilation)—April 1989—(translated from Hindi). Also, *Jawaharlal Nehru's Speeches*, Publication Division, Government of India.

Jawaharlal Nehru again said at the dedication of the project to the nation on 22 October 1963: 'This dam has been built up with the unrelenting toil of man for the benefit of mankind and therefore is worthy of worship. May you call it a Temple or a Gurdwara or a Mosque, it inspires our admiration and reverence'.

Can anyone deny that a place where thousands of people ceaselessly toiled for the good of humankind is worthy of adulation? Is this religious awe and reverence from one who was considered an agnostic, who had no religious belief, or one who was committed to the toiling Indian people?

As noted earlier, the Bhakra-Nangal Project comprises two essential parts—the Nangal barrage and connected canal systems, and the Bhakra dam and powerhouses. It would be convenient to describe the project features in terms of these two major components.

THE NANGAL PROJECT

About 13 km (8 miles) downstream of the Bhakra dam site, the Sutlej river emerges out of the Siwalik hills at Nangal, to enter the plains of Punjab. The barrage is built at this point. It is a 291 m (955 ft) long concrete structure, 27.7 m (91 ft) high and has a small storage of 30 million cu.m (24,000 acre ft), equivalent to just one day's supply in the hydel canal, to take care of the diurnal variations of releases from the dam upstream.

The Nangal hydel channel (NHC) emerges from the left bank of the Sutlej, just upstream of the Nangal barrage. It carries a discharge, beyond the silt ejectors, of 354 cumec (12,500 cusec) and serves as the feeder for the Bhakra canal system below Rupar and to generate power at the power stations at Ganguwal and Kotla at the nineteenth and thirtieth km respectively, from the head of the channel. The canal is 64 km (30 miles) long and is fully lined with cement concrete and tiles throughout its length. It runs along broken country and is crossed by as many as 58 hill torrents.

The BNP, in respect of the canal system, comprises the following:
(i) The newly constructed Bhakra canal fed by the NHC at Rupar;
(ii) remodelling of the Rupar headworks and enlarging the capacity of Sirhind canal by over 99 cumec (3500 cusec); and
(iii) the new Bist-Doab canal taking off on the right bank of the Sutlej at Rupar, with a head discharge of 39.6 cumec (1400 cusec).

The irrigation system comprises some 1110 km of main and branch canals and 3379 km of distributary channels. The whole system utilizes a discharge of 510 cumec (18,000 cusec).

Work on the barrage was completed by 1952. Even though construction of the canal system was completed only in 1954, partial irrigation commenced from 1952 itself. The total area benefitted was 4 million ha (10 million acres), of which the new area covered was 2.4 million ha (six million acres).

The Ganguwal and Kotla powerhouses have installed capacities of 77.65 MW each. They commenced operation in 1955 and 1956 respectively.

THE BHAKRA DAM

The Bhakra dam is a straight gravity concrete dam across the Sutlej river. The maximum height over the deepest foundation level is 225.5 m (740 ft), while the height over the mean river-bed level is 213.6 m (700 ft). The length of the dam at its crest is 518.16 m (1700 ft). The full reservoir level was finally kept at 515.11 m (1680 ft). The gross storage capacity is 9.62 BCM (7.80 MAF) and the effective storage is 7.19 BCM (5.83 MAF). The water spread of the reservoir is 168.35 sq. km (65 sq. miles).

There are two powerhouses. The right bank powerhouse originally had an installed capacity of 660 MW and the left bank powerhouse, 450 MW. These were later increased to 785 MW and 540 MW, respectively.

The construction commenced in 1948 and was completed in 1963. Prime Minister Nehru dedicated the Bhakra dam to the nation in October 1963.

The work on the dam was done departmentally, without using contractors. This mode of construction provided great flexibility in adjusting the designs, construction technique, and programme to suit changing needs in the field and foundation conditions from time to time. The economy, efficiency and speed of execution of the construction work amply justified the wisdom of the decision taken.[3]

The total cost of the BNP was Rs 245.28 crores. The salient features of the project are given in Appendix 5.1.

Construction

Apart from the many articles that appeared during the period of construction, several comprehensive accounts on the project had also been published. Jawaharlal Nehru had said that 'Bhakra Nangal was like a big University where we can work and while working learn so that we may do bigger things'. The next best will be to study the detailed accounts of past works of importance.

[3] CBI&P Publication No. 76, *Development of Irrigation in India*, 1965, p. 126.

Those who are interested in the intricacies and details of the planning and construction of the different components of the project should refer to the many published sources. For instance, the *Indian Journal of Power and River Valley Development* brought out a special 'Bhakra-Nangal Number' in December 1953. Similarly, the BBMB itself published the *History of Bhakra Nangal Project* in October 1988. There are many memoirs and personal accounts published by those who were connected with the project in many ways. Many professional papers and specialized articles have also been published. All these could serve as basic research material for those interested in technical, professional or detailed information. However, for the purpose of the common reader, a brief account of the project construction is given here.

NANGAL BARRAGE

The Nangal barrage was planned to be constructed by dividing the working area into two halves by enclosing the left half with a cofferdam. After this portion was concreted up to the upstream floor level, the river was diverted over the completed portion. Work was then started on the right side by enclosing it with a cofferdam.

The excavation of the foundation was restarted (after the Partition disturbances) in January 1948, using dumpers and manual labour. In April 1948, the concreting commenced and the left half was completed by March 1949. The whole river was diverted to the left side before September 1949. Excavation on the right side commenced in November 1949. The foundation concreting on the right side was over by June 1950. By February 1951, the right half of Nangal barrage (bays 13–26) was completed and the river flow was rediverted on it. The construction of the road bridge was over by the end of 1951. On 22 August 1951, within a few days of the completion of work in the left half, the Nangal barrage safely passed a record flood of 310,000 cusec (8778 cumec).

The Punjab Irrigation Department carried out the entire design work. The Nangal barrage and appurtenant structures are mainly constructed of plain or reinforced cement concrete, as appeared appropriate.

The NHC is a lined canal that begins upstream of Nangal barrage and carries water to the two powerhouses on the channel, located at Ganguwal and Kotla respectively. It thereafter feeds the Bhakra irrigation system too. As noted earlier, a special feature of this channel is that it is crossed by many hill torrents carrying high discharges.

The Ganguwal and Kotla powerhouses on the NHC have an installed capacity of 77.65 MW each.

BHAKRA DAM

The Bhakra dam is located in the Bhakra gorge. To facilitate the excavation of the foundation and construction of the dam, the Sutlej river had to be diverted from its flowing course. This diversion for construction purposes was achieved through two concrete-lined tunnels, one on either side of the river and each 15.24 m (50 ft) dia. They were nearly 805 m (half a mile) long each. The right diversion tunnel was completed by October 1953. The left diversion tunnel was ready by May 1954. The river diversion was undertaken in September/October 1954. Rockfill cofferdams were constructed upstream and downstream of the main dam foundation to dry up the area for construction. These cofferdams were big structures by themselves. They were 65.53 m (215 ft) high on the upstream and 39.62 M. (130 ft) high on the downstream side. They were started in October 1954 and completed by March 1955. Excavation of the foundation and removal of river-bed material was done using a fleet of earth-moving machinery. The excavation in the abutments in the higher elevations was started before the cofferdams were built.

In 1944, Dr Savage of the United States Bureau of Reclamation (USBR) had examined the dam site and considered it suitable. He chalked out a programme for foundation exploration. This was carried out during 1945–7 under the supervision of Dr Nickel, an engineering geologist. Extensive exploration with core drilling and drifts had, thus, been undertaken over many years. The geology of the site was also examined by a succession of eminent geologists including Dr Wiley, Mr Auden, and Dr Nickell.

The presence of clay, sheared and shattered zones, noticed during exploration and excavation, necessitated systematic foundation treatment. Consolidation grouting was done for sealing cracks and crevices. Curtain grouting was also done to provide an impervious cutoff below the foundation of the dam.

More than 3.8 million cubic metres (5 million cubic yards) of concrete were placed for the construction of the Bhakra dam and appurtenant works. Highly mechanized methods were adopted for placement of the concrete, involving a network of trestles and tracks and transportation cars and overhead cranes. In view of the great height of the dam, concrete batching, mixing and conveyance and placement was done in two stages.

Prime Minister Nehru placed the first bucket of concrete (in the spillway apron) on 17 November 1955. The first stage concreting in the dam commenced in January 1956. After the shifting of the batching and mixing plants to higher locations, the second stage concreting started in October 1958. Concreting in the main dam was practically completed by October

1961, except for the spillway bridge piers and towers and the visitors' balcony. These were completed by October 1963.

There are two power plants at Bhakra. Power Plant-I on the left bank has an installed capacity of 5 units of 90 MW each. The first bucket of concrete was placed in the foundations of the left bank powerhouse in March 1957. The concreting of the powerhouse building was done in the usual manner, in two stages. In the first stage, the foundations and the basic structure including draft tubes, scroll cases and embedded parts, were completed. By January 1959, the erection of turbines and electrical equipment commenced. In the second stage, all the remaining works were completed. The first power unit was commissioned on 14 November 1960.

Work on Power Plant-II on the right bank commenced only later in 1963 and was completed in 1968.

EXPATRIATE EXPERTS

No account of the design and construction of the project can be complete without according due recognition to a small number of expatriate specialists and experts in different fields, whose services were utilized. M.H. Slocum is recognized as an important expert of this type. He came to Bhakra from the USA on a ten-year contract signed in March 1952. His unique experience of the construction of seventeen dams in the USA and his in-depth knowledge of plants and equipment are remembered even today. B.M. Johnson headed the design directorate in the beginning. There were a number of such specialists who came for short periods and helped the Indian officers in developing expertise in various aspects of water resources development schemes.

THE INDIAN PIONEERS

The pioneering work of Indian engineers and other specialists in successfully completing the project after overcoming obstacles along the way is laudable. It is a long list indeed and includes many unsung warriors who contributed with many innovations and adaptations to suit Indian conditions. Any listing of them is bound to include the names of stalwarts like A.N. Khosla, Kanwar Sain, S.D. Khungar, M.R. Chopra, R.L. Khanna, C.L. Handa, R.S. Gill, B.R. Palta, S.C. Katoch and others. A full listing has not been attempted, lest some very deserving names get omitted by oversight. The thousands of artisans, skilled labour and workers too deserve equal praise. A large number of them lost their precious lives and yet others suffered from disabilities during the construction. Their sacrifices cannot be forgotten.

WORKSHOP

A number of job facilities were set up for the execution of the BNP. A base workshop was started at the Nangal township, with separate sections for structural fabrication, carpentry, foundry work, smithy, welding, electrical repairs, and transport. This gradually grew to be the leading complex of its kind in northern India. It now has added sections dealing with instrument repairs, penstock fabrication, fitting, and millwright shops, etc. Subsidiary field workshops at Neila near the dam site and other places were started and operated during the course of construction, as and when found necessary.

Since the completion of the Bhakra dam, the workshop has met the requirements of the Beas dam, the Bhaba hydro, Trisuli hydel and many other similar projects. It has become well known for its specialization in the fabrication and erection of penstocks for hydropower projects.

INSTRUMENTATION

A comprehensive instrumentation programme was undertaken as part of the project with a view to collect data for the study of the structural behaviour of the Bhakra dam. This comprised the installation of measuring devices for recording continuous information relating to stresses, strains, temperatures, joint movements, air and water pressure in the river outlets and the penstocks, uplift pressure, deflection, and settlements.

Seismic observatories were also installed at Bhakra by the IMD. A class I seismological observatory, two tiltmeter observation stations and two seismological observatories were set up.

Displacement, Resettlement, and Rehabilitation

The reservoir formed by the Bhakra dam has been named Gobind Sagar, in memory of Guru Gobind Singh, the great Sikh religious leader who lived in this area for many years. The reservoir covers a maximum area of 168.35 sq. km (65 sq. miles or 41,600 acres) when full. It extends to about a 100 km (60 miles) from Bhakra.

The project acquired land to the extent of 17,984 ha (44,440 acres) in the districts of Kangra (presently Una), Bilaspur, Mandi, and Solan of Punjab and Himachal Pradesh. In addition to the small town of Bilaspur, 375 villages were involved. Forty-eight villages of Kangra and fourteen villages of Bilaspur were completely submerged, while the remaining villages were affected in varying degrees.

Of the total land required, 6844 ha (16,924 acres) were already government-owned and only the balance 11,135 ha (27,516 acres) of privately owned land had to be acquired. The land was acquired as per the provisions of the Land Acquisition Act, 1894, as it then existed. This involved the migration of 7209 families, or a population of 36,000 people. The salient details are given in Table 5.1.

Table 5.1: Details of land acquisition for Bhakra submergence area

District	No. of villages affected	Total land acquired (in ha)	Privately owned land (in ha)	No. of private land-owning families
Kangra (Una)	110	5483	5483	3333
Bilaspur	256	12,313	5611	3838
Mandi	5	162	15	35
Solan	4	26	26	3
Total	375	17,984	11,135	7209

Source: BBMB.

For the urban displaced, a new town of Bilaspur was built just 2 km away on the high land overlooking the old town and the thousand urban families resettled there.

Since the number of persons affected was large, the Bhakra Control Board set up the Bhakra Rehabilitation Committee under the chairmanship of the secretary to the government, Public Works Department (PWD), Buildings and Roads, Capital Project, Chandigarh. It was asked to advise the government on the following matters:

(i) • Principles and methods of rehabilitation with particular reference, to—basis of rehabilitation *vis-à-vis* land for land, cash compensation etc.

• places of resettlement—after ascertaining public opinion—both among the population to be displaced and among the people of the area where the displaced persons would be rehabilitated.

• fixing responsibilities of the government/authorities for rehabilitation.

(ii) Procedure for determining compensation to the displaced persons.

(iii) Procedure for determining compensation in individual cases.

(iv) Rough cost estimates and recommendations regarding its incidence.

(v) Construction of new town in lieu of Bilaspur.

The general manager, Bhakra dam, joint secretary, Revenue, Himachal Pradesh, deputy commissioner, Bilaspur and deputy commissioner, Re-settlement, were the official members. The three members of the Himachal Pradesh Territorial Council and the member representing Kangra in the Legislative Assembly were the non-official members.

It was decided that in respect of the lands submerged, the displaced persons might be compensated, as far as possible, in the form of land. In the case of those who did not want land in lieu of cash compensation, they could be paid in cash or partly in land and partly cash. Compensation for houses, trees, etc., was in cash. The compensation was awarded by the land acquisition collectors concerned under the Land Acquisition Act. Liberal compensation was paid for land, houses, trees, *gharats*, and other property going under submergence. The acquired lands were even leased out to the erstwhile landowners temporarily till the actual submergence, with the proviso that they could not evict existing tenants. The owners of houses were permitted to make free use of material from their houses, regardless of the acquisition value.

RESETTLEMENT

As per the resettlement policy, the stakeholders were also consulted and their views fully considered. Nearly a third of the total affected families were resettled in nearby areas of Himachal Pradesh and a little over a third of the families preferred to receive cash compensation so that they would resettle on their own. A little less than a third of the families desired to resettle in land within the irrigation command of the project. The details are given in Table 5.2.

Table 5.2: Rehabilitation of Displaced Families

Category	Number of families
1. Resettled within Himachal Pradesh by the Himachal Pradesh Government	2398
2. Those who preferred cash compensation and resettlement on their own	2632
3. Families that preferred resettlement in the irrigation command	2179
4. Total	7209

Source: BBMB and Dy. Commissioner, Resettlement, BNP.

In addition to resettlement, the following actions were taken to provide means of livelihood and more amenities for the families resettled in Himachal Pradesh:

- Free fishing licenses in the reservoir for three years.
- Gainful employment to local people on the construction of the project.

Where a small area of land or immovable property was still left unacquired and the landowners desired its acquisition, that too was permitted.

Surveys for locating suitable land in the Sirmour, Mandi, and Mahasu districts of Himachal Pradesh, as also in the Hoshiarpur and Ambala districts of Punjab, were carried out but very little land was available there. However, a similar survey in the Hisar district showed that sufficient land in compact blocks at cheap rates was available. The displaced people too expressed their preference for lands coming within the irrigation command of the BNP. About 5342 ha (13,200 acres) of land in the BNP irrigation command were acquired in compact blocks in thirty villages of the Hisar district (since trifurcated into Hisar, Sirsa, and Fatehabad districts).

Compared to the price paid for the land that was to be submerged, the cost of land to be newly acquired in Hisar was low. Thus, it was found that the full value of compensation for submergence land could not be made in the form of only land in the command. Therefore, it was decided that compensation would be partly in the form of land and partly in cash, subject to two overriding considerations. These were that no oustee would get more than 25 acres (about 10 ha) of land and that no oustee would get less than his cultivated land acquired for submergence, if the compensation amount was adequate to meet its cost.

In order to help small landowners among the oustees, compensation up to Rs 1000 was made fully in the form of land. Beyond that, a system of graded cuts that worked on a slab system was followed for additional land.

The details are given in Appendix 5.2 and the list of the thirty villages is in Appendix 5.3.

Landless tenants in the submerged area were also declared eligible for allotment of land in the command. They were given land to the extent of their tenancy (as recorded in revenue records for the 1957 *kharif* season), subject to a maximum of 5 acres (about 2 ha), the price to be paid by them in twenty equal half-yearly instalments, commencing after a grace period of two years.

Even artisans and labourers affected by the project, who did not own or cultivate land as tenants but wished to shift and settle in the Hisar district, were each allotted a half acre of land, free of cost.

Allotments of *abadi* plots for the construction of residences and shops in the resettlement villages were also done. Abadi sites were on the basis of four allottees per acre. Model layouts for abadi sites were planned in each resettlement village.

The following additional amenities were provided:

- temporary shelter accommodation;
- easy loans for construction of houses and supply and transport of building materials;
- repairs to old wells and new wells for drinking water supply, where necessary;
- Bhakra canal water supply through new canals, minors, and water courses;
- bridges on canals, new ferry services, new roads/village paths;
- new primary schools, medical facilities, cattle inoculation, and security arrangements; and
- rail/bus fare plus a lump sum rehabilitation grant.

In order to resettle the oustees as near to their old *biradari* (social and cultural group setting) of neighbouring villages as far as possible, allotment of land in the new area was made on the basis of *had bast* numbers in the erstwhile villages in Kangra, Bilaspur, and Mandi.

Deeds of conveyance of proprietary rights to the allottees were executed after they fulfilled all the conditions of their allotment and clearance of all pending recoveries of dues from them.

FACILITIES

The public health unit of the project was set up in 1947. Anti-malarial measures including DDT spraying, vaccination, cholera and typhoid inoculation, etc. were all made part of its work. A fifty-bed hospital was set up by the project in 1951 and this was gradually expanded.

Nangal was provided with a unit of the Punjab Red Cross Society. Maternity, childcare and other facilities became available from the very beginning of the project.

WATER SUPPLY

A regular water supply system was started at Nangal in 1947. Both raw water for lawns and gardens, and treated potable water for the people were supplied. A similar local water supply for Bhakra and other places was also provided.

SCHOOLS

A number of schools for boys and girls were opened in the Nangal township project colony for the education of children of employees as well as the local population. They were later transferred to the education department of the Punjab government.

The project improved the position of the drinking water supply in a large number of villages and towns, not only in the command area and nearby but even in some far-off places. The metropolitan capital Delhi, as well as the cities of Chandigarh, Patiala, Ropar, etc., are some examples.

A number of factories depend on the project for their water supply. Some major examples are the National Fertilizers, Nangal, Thermal Power Station, Delhi, Ropar Thermal Power Station, etc. A large number of small industrial units are also similarly benefitted.

PROACTIVE PLANNING BENEFITS FOR INDIA

The Government of India had made contingent plans to fully harness the waters of the Indus system likely to be allocated to it, and kept this updated as the treaty negotiations were progressing. All the participating states agreed, at a special meeting of the Bhakra Control Board held in August 1951, on the extent of areas to be included within the irrigation boundary of the project, as well as the water allowance and capacity factors. Soon after the World Bank made its suggestions for the Indus solution, the states concerned within India held detailed discussions. They reached an agreement in 1955 on their respective shares in the beneficial use of the waters of all the 'eastern rivers'. The Government of India assisted them in this process and encouraged them to make detailed project plans. This proactive measure enabled the early development and use of the waters by India.

Indian engineers and planners should be given full credit for having enabled optimal use of the Sutlej waters allocated to India, without any loss of time in difficult, dispute-ridden circumstances. They were able to proceed with the planning and construction of the BNP and other interrelated projects, using the waters of the Indus system—a feat that deserves recognition and praise. This planning envisaged the full utilization of the flows in the Sutlej after the Bhakra dam was completed which, too, was achieved in the shortest possible time.

INTEGRATED OPERATION OF SUTLEJ AND RAVI-BEAS SYSTEMS

After the Bhakra canal was opened for kharif irrigation in 1964, the Sutlej water during the kharif period was expected to fall short of needs during critical periods. After the sharing of the surplus Ravi-Beas waters was also agreed upon by the states concerned, an integrated operation of the Sutlej and Ravi-Beas systems was conceived. A portion of the Sirhind canal area in Punjab and new areas in Rajasthan were recommended to be irrigated by the Sirhind feeder, which emerged lower down from the Harike barrage.

Consequent to the execution of Unit I of the Beas project, a part of the Beas waters was diverted to the Sutlej and an integrated operation worked out. This scheme involved a Beas-Sutlej link diverting 4.7 BCM (3.8 MAC ft) of Beas waters at Pandoh to the Sutlej at Dehar. This enabled the generation of hydropower at Dehar, additional hydropower generation at the Bhakra dam, as also extension of irrigation to 0.4 million ha. This part came into operation in 1980.

ORGANIZATIONAL SET-UP

With a view to ensuring the efficient, economical and early execution of the project and to maintain control on the construction activities, as also to bring together the representatives of the partner states and the Government of India, the Bhakra Control Board (BCB) was set up in September 1950 and chaired by the Governor of Punjab. The secretary to the Government of India, Ministry of Irrigation and Power, was the Vice Chairman. The Board met at regular intervals and took decisions on all matters relating to the project expeditiously. The construction work was placed under a general manager. A chief accounts officer took care of all accounts matters. To advise the Government on various engineering and planning problems that arose during the design and construction phases, a Bhakra Board of Consultants was also set up. When the Punjab state was reorganized, the Bhakra Management Board (BMB) was set up under the Punjab Reorganization Act of 1966 and was later renamed the BBMB when the components of the Beas project were transferred to it.

REGULATION OF WATERS

The BBMB maintains and operates the Bhakra-Nangal and Beas projects on behalf of the participating states of Punjab, Haryana, Rajasthan, and Himachal Pradesh. It has been made responsible for the regulation of the waters of the three 'eastern rivers' namely, the Ravi, the Beas, and the Sutlej and passing on allocated shares of the waters to the states concerned at predetermined points. This involves, *inter alia*, the operation of multipurpose reservoirs including the Bhakra and Beas in an integrated manner. The process of regulation involves the accumulation and assimilation of the decision support data, decision making and implementation thereof on a day-to-day basis. It is, indeed, an intricate exercise, particularly in view of possible conflicting demands between the states and different functions.

Many large dams serve a number of purposes with a single facility. In fact, the National Water Policy of India requires that water resource development projects should, as far as possible, be planned and developed as multi-purpose projects. However, under a multi-purpose reservoir the interests of the various components such as irrigation, power generation, and flood control are often at variance with one another, even in cases where these are owned and operated by a single state/authority. Such conflicts become pronounced when more states/authorities are involved. In an ideal case, the best manner of regulating the flows would imply an enlightened compromise of these relative interests/states.

The potential for conflict exists between the flood control objective for the reservoir operation (where storage space for absorbing the incoming flood peak is needed) and hydropower or irrigation (where it might be considered desirable to keep as high a storage as possible in order to meet later demands). Moreover, the requirements of water releases to meet irrigation demands over different sub-periods may not synchronize with the needs of hydropower generation. Water releases through turbines at peak power demand hours are rarely suited for irrigation needs. A well-laid out reservoir regulation manual could obviate many difficulties but the potential for conflicts in day-to-day regulation surely exists.

In the case of the Bhakra reservoir, the inflows are mainly from the Sutlej and its tributaries. The waters of the Beas diverted at the Pandoh dam for power generation at Dehar also flow into the Sutlej and contribute to the inflows.

The Bhakra reservoir operations could be classified into two distinct parts of the hydrological year, namely the filling period and the depletion period. The filling period broadly covers the monsoon months when kharif crops are sown and is counted from 21 May to 20 September. Similarly, the depletion period for the Bhakra reservoir is considered to be from 21 September to the following 20 May.

The erstwhile state of Punjab (including Haryana) and Rajasthan entered into the Bhakra-Nangal Agreement in 1959, which governs the respective shares in the waters of the Sutlej accorded to them. The BBMB has set the upper limits to which the reservoir could be stored on different benchmark dates in order to take care of flood management needs too.

The manner of the depletion of the reservoir is planned each year towards the end of the filling period after taking into account the storage position and the requirements of water for *rabi* crops. Some carryover is also planned to cater for likely poor inflows during the next filling season, particularly the early part. The partner states intimate their projected rabi irrigation requirements. Based on these projections and within their entitlements, detailed waterpower studies are made and updated each month. These are used to arrive at the real-time operation of the Bhakra reservoir.

The BBMB has constituted a committee presided over by its chairman and comprising its wholetime members, chief engineers for irrigation of the partner states and technical members of partner electricity boards as members. This committee takes stock of all relevant data, and decides on the supplies to be delivered to the partner states at the contact points and the consequent releases from the reservoir every ten days. The regulation of actual releases each day also takes into account the actual field conditions.

An index map of the project and its canal system is provided in Figure 5.1 and the typical spillway section is shown in Figure 5.2.

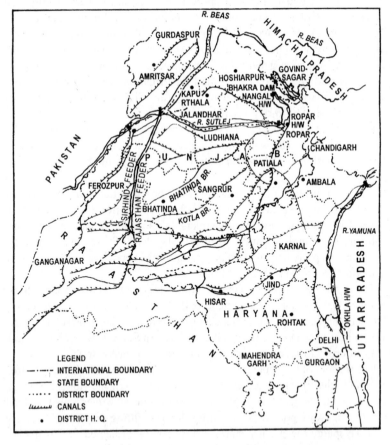

Figure 5.1: Index Map (Bhakra-Nangal Project)

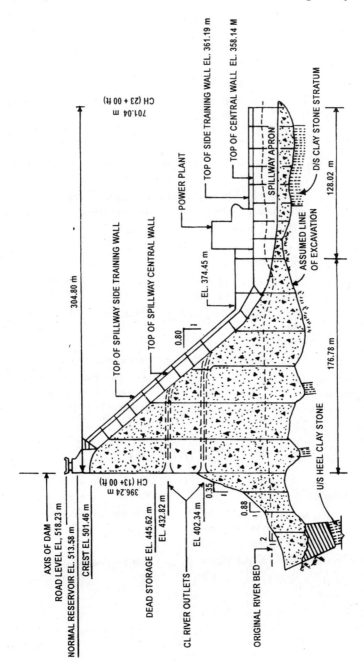

Figure 5.2: Bhakra dam (Maximum Spillway Section)

APPENDIX-5.1

Bhakra-Nangal Project: Salient Features

Details	Bhakra dam		Nangal barrage	
	In metric units	In Foot Pound System (FPS) units	In metric units	In FPS units
1	2	3	4	5
A. General				
Location	Dam near the village Bhakra in Himachal Pradesh		A barrage 13 km (8 miles) downstream of the Bhakra dam	
River	Sutlej		Sutlej	
Purpose	Irrigation, hydroelectric power generation and incidental flood control			
Catchment area	56,980 sq. km.	22,000 sq. miles	Practically the same as at Bhakra	
Average annual run-off	16.775 BCM (1911–57 series)	13.60 M.ac.ft	Practically the same as at Bhakra	
Design flood	11,327 cumec	400,000 cusec	9912 cumec	350,000 cusec
Commencement of construction	1948		December 1946, but stopped soon due to communal riots Restarted in 1948	
Year of completion	1963		1952	
Partial benefits from the year	1958		1954	
B. Main Structure				
Type	Concrete straight gravity dam with radial gate overflow spillway		Concrete barrage	
Bed rock	Sandstone with claystone		Massive conglomerate	
Max. height at deep foundation	225.55 m	740 ft	29 m	95 ft
Height above river bed	167.64 m	550 ft		

(Contd)

Appendix 5.1 (contd)

1	2	3	4	5
Length at top	518.16 m	1700 ft	291 m	955 ft
Bottom length	99 m	325 ft	291 m	955 ft
Width at top	9.14 m	30 ft	Road; bridge width 6.71m or 22 ft	
Width at base	190.5 m	625 ft		
Elevation of top of dam	518.16 m	1700 ft	Max. pond level	RL 1154 ft
			Top of breast wall	RL 1160 ft
Spillway type	Centrally located overflow type			
Design max. discharge gates	5587 cumec	197,300 cusec	9912 cumec	350,000 cusec
Type & no. of spillway gates	4 radial gates			
Size of gates	15.24 × 14.48 m	50 ft × 47.5 ft		
Spillway crest	501.45 m	1645.21 ft		

River outlets and flood control gates

No. of outlets	16 in 2 tiers of 8 each at El. 402.33m (1320ft*) & 432.80m (1420ft*)			
Size of outlets	2.64m × 2.64 m	8.67ft × 8.67 ft		
Outlet Discharge	3002 cumec	1.06 lakh cusec		
Spillway & outlet discharge	8212 cumec	2.9 lakh cusec		
Discharge if reservoir level goes up to maximum reservoir level (MRL)	11,328 cumec	4 lakh cusec		

Provisions against seismic forces

Maximum horizontal earthquake acceleration of 0.15 g with period of vibration of 1 second and direction of vibration normal to axis of dam provided for. Similarly, maximum vertical EQ acceleration of 0.075 g acting upwards with period of vibration of 1 second provided for.

(Contd)

Appendix 5.1 (contd)

1	2	3	4	5
Bhakra reservoir (Gobind Sagar) details			Nangal pond	
MRL	515.11 m	1690 ft*	270 m	670 ft
Normal high reservoir level	513.58 m	1685 ft*		
Full reservoir level (FRL)	512.06 m	1680 ft	Not applicable	
Dead storage level	445.62 m	1462 ft		
Maximum reservoir area	168.35 sq. km.	65 sq. miles		
Gross storage at high reservoir level	9.62 BCM at 513.58 m level	7.80 M.ac.ft at 1685 ft level	30 MCM	24,000 acre ft
Corresponding live storage	7.19 BCM	5.83 M.ac.ft		
Gross storage at restricted level	9.337 BCM	7.57 M.ac.ft		
Corresponding live storage	6.907 BCM	5.6 M.ac.ft	Not applicable	
Dead storage	2.43 BCM	1.97 M.ac.ft		
Submergence	Bilaspur town and 375 villages			

Displaced families

7209 families or 36,000 people were displaced by the project. Of these 2179 families opted for land compensation and were allotted land in the command area. People displaced from Bilaspur town (4000) rehabilitated in new Bilaspur town.

(Contd)

* Maximum permissible reservoir level restricted to 1680 ft (except for flood management) since 1990.

Appendix 5.1 (contd)

Irrigation—metric/FPS systems—[million hectares (m.ha.) & million acres (m.ac.)]

Category/ State	Punjab post-1966	Haryana post-1966	Punjab & Haryana (combined)	Rajasthan	Total
Gross area	0.97 m.ha or 2.39m.ac.	1.27m.ha. or 3.15 m.ac.	2.24 m.ha. or 5.54m. ac.	0.44 m. ha or 1.08 m. ac.	2.68 m.ha or 6.62 m.ac.
Culturable area	0.86m.ha or 2.13 m.ac.	1.14 m.ha. or 2.81 m.ac.	2.00 m.ha. or 4.94 m.ac.	0.37 m.ha or 0.92 m.ac.	2.37 m.ha or 5.86 m. ac.
Irrigation area	0.39 m.ha or 0.96 m.ac.	0.68 m.ha or 1.67 m.ac.	1.07 m.ha or 2.63m. ac.	0.23 m. ha or 0.57 m. ac	1.30 m.ha or 3.20 m. ac.

Summary—Hydropower installed capacity (in MW)

Powerhouse	Number of units	Capacity of each unit	Originally installed total	Years of commission	Present capacity
Bhakra dam— Left bank	5	90 MW each	450 MW	1960–61	540 MW (5 × 108 MW)
Bhakra dam— Right bank	5	120 MW each	600 MW	1966–68	785 MW (5 × 157 mw)
Ganguwal (on hydel channel)	3	2 of 24.2 MW and 1 of 29.25 MW	77.65 MW	1955–62	77.65 MW
Kotla (on hydel channel)	3	2 of 24.2 MW and 1 of 29.25 MW	77 .65 MW	1955–61	77.65 MW
Grand total installed capacity		Original	1205.3 MW	As of June 2003	1480.30 MW

COSTS AND BENEFITS

COSTS

	Rs in crores (rounded off)
Bhakra dam	52.5
Bhakra dam powerhouses—right and left	98.3
Nangal barrage	3.2

(Contd)

NHC	10.2
Powerhouses on NHC (two)	13.7
Remodelling Rupar headwork(s) and Sirhind canal	5.9
Bhakra canals	43.1
Transmission lines	18.4
Total cost	Rs 245.3 crores

BENEFITS

A. *As per original BNP*

Area newly irrigated   *2.372 million ha Cultural Command Area (CCA)*

State	CCA	Irrigation area (both in million ha)
Punjab	0.863	0.389
Haryana	1.137	0.676
Rajasthan	0.372	0.231
Total	2.372	1.291

B. *After integrating Beas flows*

New area irrigated	*2.6 million ha (6.5 million acres)*
Area where irrigation is improved	0.9 million ha (2.2 million acres)
Number of towns electrified	128
Number of villages electrified	13,000
Flood moderation	

Industrial and Domestic Water Supply

Fisheries, industrial development, tourism, etc.

APPENDIX-5.2

Scheme for allotment of land to Bhakra Oustees in Hisar district

Amount of compensation	Compensation to be paid in cash	Compensation to be paid in the form of land	Remarks
1	2	3	4
Up to Rs 1000	NIL	100%	Full allotment of land
Above Rs 1000 Up to Rs 2000	60%	40%	Every allottee will get land for the basic value of Rs 1000 and a cut of 60% will apply on the amount of his compensation which will be over Rs 1000
Above Rs 2000 Up to 3000	65%	35%	Every allottee will get land for the basic value of Rs 1400, and a cut of 65% will apply on the amount of his compensation which will be over Rs 2000
Above Rs 3000 Up to 4000	70%	30%	Every allottee will get land for the basic value of Rs 1750 and a cut of 70% will apply on the amount of his compensation which will be over Rs 3000
Above Rs 4000 Up to 5000	75%	25%	Every allottee will get land for the basic value of Rs 2050, and a cut of 75% will apply on the amount of his compensation which will be over Rs 4000
Above Rs 5000 Up to 6000	80%	20%	Every allottee will get land for the basic value of Rs 2500, and a cut of 80% will apply on the amount of his compensation which will be over Rs 5000
Above Rs 6000 Up to 7000	85%	15%	Every allottee will get land for the basic value of Rs 2500, and a cut of 85% will apply on the amount of his compensation which will be over Rs 6000
Above Rs 7000 Up to 9000	90%	10%	Every allottee will get land for the basic value of Rs 2650, and a cut of 90% will

(Contd)

Appendix 5.2 (contd)

1	2	3	4
Amount of compensation	Compensation to be paid in cash	Compensation to be paid in the form of land	Remarks
			apply on the amount of his compensation which will be over Rs 7000
Above Rs 9000	95%	5%	Every allottee will get land for the basic value of Rs 2850, and a cut of 95% will apply on the amount of his compensation which will be over Rs 9000

(i) No oustee will be allotted more than 25 acres.

(ii) No oustee will be allotted land less than his cultivated area, subject to a maximum of 25 acres, provided his compensation amount is adequate to meet the cost.

(iii) When any money becomes payable by the oustee by court decision or any reason whatsoever, the oustee will be required to pay the same within reasonable time, failing which his allotment will be liable to be cancelled, and the government will have further right to use the allotted land in such manner as it thinks fit.

(iv) Where government evolves a model scheme for the construction of village houses, the allottee shall be bound to abide by the conditions of the model scheme relating to the manner in which a house may be constructed.

(v) Deed of conveyance of proprietary right to the oustee will only be executed when the oustee has fulfilled all the conditions of settlement and paid all the sums which may be due from him.

Note:

1. The cut shown against each grade will apply to that portion which falls within a grade, not to the total holding.

2. If as a result of the application of this sliding scale, the allotment of land to an oustee works out to less than five acres then no cut is to be imposed.

APPENDIX-5.3

List of 30 Villages in Hisar

1. Rawalwas Khurd
2. Gajuwala
3. Hyderwala
4. Chillewal
5. Nanheri
6. Kunal
7. Nathwan
8. Burj
9. Ayalki
10. Razabad
11. Kata Kheri
12. Boswal
13. Bhirrana
14. Bhutan Khurd
15. Gilla Khera
16. Hijrawan Khurd
17. Jodhka
18. Ram Nagaria
19. Bhamboor
20. Liwalwali
21. Baruwali-II
22. Elenabad
23. Abholi
24. Jhornalli
25. Odhan
26. Ram Nagar
27. Gobindgarh
28. Desu Jodha
29. Dhanoor
30. Jamalpur Shaikhan

6

IRRIGATION FROM THE
BHAKRA-NANGAL PROJECT

The Project is Reoriented

The agricultural policy in India was greatly influenced and affected by the acute food shortage that occurred during the Second World War years. Even prior to its outbreak, the average annual net cereal import (for the three years ending 1938–39) was over two million tonnes. It has already been noted that, for its own reasons, the energy of the pre-Independent Government of India, for many decades, was concentrated in developing the agricultural prosperity of the western part of the Punjab. There was an apparent neglect of the southern and eastern parts of united Punjab, particularly in respect of irrigation and other infrastructure relating to agriculture. The effect of this skewed development emphasis became apparent in August 1947, with the partition of India. West Punjab, which had an annual food grain surplus of over a million tonnes became part of Pakistan. The entire economy of Punjab was completely upset overnight, not to speak of its serious impact on India as a whole. The thirteen districts that constituted the newly carved-out province of East Punjab were the worst affected due to Partition and its aftermath. Practically the entire area growing American cotton was lost and textile mills had to import cotton for survival. The truncated part of Punjab that remained in India was instantly faced with acute shortages in agricultural products, as also in industrial outputs.

The Bhakra project was reoriented and planned in order to provide urgently needed irrigation facilities to long-neglected arid and semi-arid tracts of lands in East Punjab and Rajasthan. The need to assure water supply to the then existing irrigation systems of the Sirhind canal, the Grey canal, and the Sarsuti and Ghaggar canals was also kept in view during the planning of the new irrigation canal system. This necessitated the remodelling of the Roopnagar (Ropar) headworks. Modifications were carried out in the

Sirhind canal system and the canal system for new irrigation command that emerged at the tail of the Nangal hydel canal near Roopnagar, as well as in the Nangal barrage and Bhakra dam. This had to be done while keeping the flows to the existing canal systems in India uninterrupted, also taking into account India's commitments and understandings in respect of the water supplies to downstream areas that now constituted Pakistan.

Existing Area—Sirhind Canal

The Ropar headworks feeding the Sirhind canal system are among the oldest developments in the Indus basin, particularly based on the flows of the Sutlej. The project was undertaken in 1873 and opened in 1882. The river finally comes out of the hills near Roopnagar, where the Sirhind canal begins on the left side of the Sutlej. At the time of Partition, the culturable command area (CCA) was 3.72 million acres (1.5 million ha) and the carrying capacity of the canal in the head region was 9040 cusec. The Sutlej generally carried large flows during the monsoon months (July to September) but, in the absence of upstream storage, these dwindled to around 3000 cusec in winter. The channels had to be run by a rotation system in three or four groups. The design of the old weir had certain limitations too, in the control over the river flows and canal supplies. Owing to the comparatively large CCA and limited winter flow, the intensity of cropping, the capacity factor and the water allowances on the system were all very low.

The irrigation canal system existed only on the left side of the Sutlej, while the areas on the right were left out. The groundwater levels on this side had depleted and the need to extend the canal system on the right side was felt too. There were also many tracts of land even within the irrigation boundary of the Sirhind canal system, which remained unirrigated, where canal irrigation needed to be extended. The old Roopnagar headworks required remodelling to take care of many existing problems and to serve new areas better after the storage at the Bhakra dam fructified. The remodelling of the Roopnagar headworks and Sirhind canal were, therefore, made part of the BNP.

The old Ropar weir was converted to a barrage and provided with new undersluices, as per modern techniques. Its various components too were remodelled. The head regulator of the Sirhind canal was remodelled and its carrying capacity increased to 12,625 cusec. A new head regulator was provided on the right side of the Sutlej river and a new canal, called the Bist-Doab canal, with a carrying capacity of 1408 cusec, emerged from it to serve new areas in Jalandhar, Hoshiarpur, and Kapurthala districts. The gross area of this tract is 2.25 million acres (0.91 million ha). The area irrigated by this canal system is 1.0 million acres (0.4 million ha).

The Sidhwan branch is a new link channel under the Bhakra canals, which links the old inundation canals, namely, Daultwah and Kingwah, that previously used to be fed directly from the Sutlej river. Before Bhakra, the Patiala feeder was providing supplies to the Kotla, Ghaggar and the Choa branches. With the commissioning of the Bhakra canals, part supplies for the Ghaggar and the Choa branches are tapped directly from the Bhakra main line through the Ghaggar link.

New Area—Bhakra-Nangal Canal System

The Government of Punjab constituted a crop planning committee in late 1952, after the outlines of the project took shape and the states had come to some understanding about the likely water availability for them, even though the final picture had yet to be perceived. The committee recommended the crop pattern tractwise, taking into consideration the climatic conditions, rainfall, water table, type of soil, local agricultural practices and various other local factors.

The areas to be served by the Bhakra canal system were broadly classified into three zones, based on a careful consideration of the different areas needing irrigation and of all relevant matters, including the climatic and physiographic characteristics detailed above.

ZONE I—RESTRICTED PERENNIAL

This zone includes the Bist-Doab canal that emerges from the Roopnagar barrage and the Samrala, Rajpura, south of Patiala and the Kaithal-Pehowa tracts. These areas are densely populated and generally receive good rains during the monsoon and some moderate rain in winter. However, the rainfall was deficient during the sowing and maturing seasons. Therefore, restricted perennial supplies for canals in this tract were proposed with a view to safeguarding against possible waterlogging and to conserve water for other needy tracts. The water allowance was fixed at 2.25 cusec per thousand acres (0.06 cumec per 404.69 ha). A flow intensity of 45 per cent was adopted for this zone. The total CCA in Punjab (including Haryana) under this zone accounted for 1.5 million acres (0.64 million ha).

ZONE II—NON PERENNIAL

This zone comprises riverine areas with high spring levels situated close to the river. Most of these areas were already receiving inundation supplies through the 'Grey canal series'.[1] The canal supplies were made available for a limited duration only. Besides, a low intensity of 35 per cent of CCA was proposed for this zone and it was proposed to keep the canal open up to the end of October to give a first watering for the rabi crops. The water allowance was kept as 3.5 cusec per thousand acres (0.1 cumec per 404.69 ha). A total CCA of 0.6 million acres (0.27 million ha) in Punjab (including Haryana) comes under this zone.

ZONE III—PERENNIAL

This zone consists of arid and semi-arid areas of Hisar/Sirsa of Haryana and the Rajasthan areas beyond. These areas receive very meagre and, moreover, uncertain rainfall, varying from 10 to 15 inches annually (25 to 38 cm). The spring levels are also generally low, varying from 100 to 150 feet (30 to 45 m) below the ground. Well water was also found to be brackish in many of these areas. This tract had been subject to frequent famines in the past. The water allowance was kept as 2.75 cusec per thousand acres (0.08 cumec per 404.89 ha). An intensity of 62 per cent was adopted. The CCA of the tracts of this zone is 3.67 million acres (1.49 million ha), of which about 0.9 million acres (0.37 million ha) lie in Rajasthan and the balance in Punjab and Haryana.

The number of days of supplies at the above-mentioned rates to be drawn from the Bhakra storage during the depletion period (21 September to 20 May) was set at 157.3 days for restricted perennial, 52.9 days for non-perennial and 173.7 days for perennial areas.

The proposed water allowances and capacity factors and the areas to be included within the irrigation boundaries of the BNP were accepted by all the participating states at a special meeting of the BCB held in August 1951 at New Delhi.

The entire area under the Sirsa branch of the Western Jamuna canal (WJC) has been diverted to be irrigated from the Narwana branch of Bhakra main line during the non-monsoon months, thus leading to a saving during that period of 1794 cusec (50.8 cumec). It was now agreed that the remaining channels of the WJC could use this for improving the capacity factors during the period of shortage.

The new Bhakra canals and the existing Sirhind canals would be operated as one system, sharing shortages/excesses in the ratio of their authorized

---

[1] Grey canals are a system of inundation canals built during British rule to serve irrigation needs. They were built by Col. I.J.H. Grey, Deputy Commissioner of Ferozpur, in 1875–6.

full supply discharges. This was agreed to at the BCB meeting held in August 1951 by all the participating states. As the climatic conditions of the tail-end areas of the Sirhind canal are similar to the Hisar and Bikaner areas, it was agreed that the water allowance to the canal would be improved to that of the perennial areas of the Bhakra system. Owing to these decisions, it was considered that the irrigation in the Sirhind canal areas would increase by 0.4 million acres (0.16 million ha).

The areas served by the Ghaggar and the Sarsuti canals, having been included under the Bhakra canal system, the flows in the Ghaggar and the Sarsuti so released were to be used by Punjab above Ambala.

The Bhakra main line starts from the tail of the Nangal hydel canal at Roopnagar. It has a designed discharge capacity of 12,500 cusec (354 cumec). It proceeds almost straight towards Tohana, situated at the border of the Hisar district. In addition to direct distributaries, aggregating to 1226 cusec (35 cumec), the major branches that emerge from it are as under:

Name of branch	Full supply discharge (cusec)
Narwana Branch	4459
Ghaggar Branch	1433
Choa Branch	314
Bhakra Main Branch	5069
Fatehabad Branch	1707
Ratia Branch	671
Rori Branch (with Ottu Feeder)	1030
Karni Singh Branch	448
Sadul Branch	1244
Barwala Branch	824

As noted earlier, the Sidhwan branch emerges from the Sirhind canal system with a 1727 cusec capacity, and the Bist-Doab canal starts from its head regulator near Roopnagar barrage with a 1401 cusec capacity.

The Bhakra main line, the main branch, and the Narwana, Karni Singh, and Sadul branches are lined channels; the rest are not. The Bhakra main and Fatehabad branches emerge from the tail of the Bhakra main line. A summary of the CCA and irrigable areas under the different states as originally approved and as they stand now after the reorganization of combined Punjab into Punjab and Haryana in November 1966 are given in Table 6.1.

The total CCA under the project is 2.372 million ha (5.86 million acres) and the irrigation area 1.296 million ha (3.20 million acres).

Table 6.1: BNP—New Command Area

(all areas given in 1000 ha)

Zones	Item	Erstwhile Punjab	Erstwhile PEPSU	Total Punjab and PEPSU	Rajasthan pre- and post-1966	Punjab Post-1966	Haryana Post-1966	Project total—all states
I	GCA	494	220	714	–	526	188	714
Restricted	CCA	445	198	643	–	473	170	643
Perennial	IA	200	89	289	–	213	76	289
II	GCA	222	48	270	–	270	–	270
Non	CCA	200	43	243	–	243	–	243
perennial	IA	70	15	85	–	85	–	85
III	GCA	1037	221	1258	438	172	1085	1695
Perennial	CCA	916	198	1114	372	147	967	1486
	IA	568	123	691	231	92	600	922
Total	GCA	1752	489	2242	438	968	1273	2679
command	CCA	1560	440	2000	372	862	1137	2372
	IA	838	228	1065	231	388	676	1296

Source: Statistical Abstracts of Punjab and Haryana.

Notes: 1. GCA = Gross command, IA = Irrigable area, CCA = Culturable command area
2. Totals may not exactly tally due to metric conversion and rounding off.

WATER SUPPLY REQUIRED

With the normal reservoir level taken as 512.06 m (1680 ft), the net utilizable storage space is 5.625 MAF. The total water supply required for irrigation, as per the agreed water allowances and capacity factors, was worked out as 8.365 MAF. However, this entire supply is not required to be given from the storage. On an average, it was considered possible to feed the canals direct from the free river flow during 21 May to 20 September each year. The net supplies required to be given from the storage would in that case be 5.255 MAF. Some additional storage would be needed for improving irrigation supplies in the Sirhind canal. The total storage requirements to meet all demands was 6.2 MAF which meant an average shortage of 0.58 MAF as a minimum. The project planners considered that this could be met from the link to the Sutlej from Ravi-Beas waters as also monsoon season flows in the Yamuna for Narwana branch (old Sirsa branch).

PROTEST BY PAKISTAN

India was getting ready to operate the Nangal canal system from the summer of 1954, limiting its withdrawals to the surpluses in the Sutlej river and to such waters as Pakistan could replace from the Chenab river through the Balloki-Suleimanke Link and the Bambanwala-Ravi-Bedian link. Pakistan, in a formal note sent on 5 June 1954, asked India to 'refrain from making any new or increased withdrawal in anticipation of an agreement that remains to be worked out'. Pakistan's ambassador in the USA had also written to the World Bank president, conveying their readiness to work out an ad-hoc arrangement for 1954 but making it clear that this 'must not be taken as implying acceptance of the division of supplies suggested by the Bank'. Prime Minister Nehru wrote on 21 June 1954, to the World Bank president that 'the persistently negative and uncooperative attitude of Pakistan has therefore made impossible the continuation of talks initiated by you in March 1952 and Pakistan has thereby voided the understanding under which we have been working for the last two years'.

Opening of the Bhakra-Nangal Canal

On 8 July 1954, Prime Minister Jawaharlal Nehru opened the canal system and dedicated 'the Bhakra-Nangal works to the good of the Indian people'. This was followed by celebrations, which continued for the next ten days, as the life-giving waters reached, for the first time, different places along the long canal.

The government, press, and political parties in Pakistan, however, showed their great indignation at the opening of the Bhakra canal by India. In a letter to Nehru, the prime minister of Pakistan, Mohammad Ali, called the opening of the Bhakra canals as a 'precipitate action' and added that it 'has struck a serious blow to the cause of Indo-Pakistan amity'. According to the *Pakistan Times*, at a press conference held on 15 July 1954, the Pakistan prime minister said that it 'was a potential threat to peace between the two countries'.

Nehru, in his reply to the Pakistan prime minister, referred him to the many notices given earlier and pointed out that 'even on the opening of the Bhakra-Nangal Canal and thereafter, we did not reduce the normal supply of irrigation water to Pakistan'.

In order to supply water during the non-monsoon months, partial storage was started much before the completion of the Bhakra dam in 1963. The maximum reservoir level during the filling season of 1961 was 472.44 m (1550 ft) which was depleted, thereafter, to 438.9 m (1440 ft) by May 1962 for supplying water to the irrigation canals. Similarly, the maximum water level reached in the following five filling seasons were:

Year	Maximum water level reached	
	In metres	In feet
1961	472.4	1550
1962	490.7	1610
1963	499.9	1640
1964	506.3	1661
1965	484.9	1591

During the initial filling stages, the dam was constantly under watch and it was found that 'the structural behaviour of the various works was excellent and in consonance with the design assumptions'.[2] It was only in 1975 that the maximum reservoir level of 514.3 m (1687.36 ft) was reached.

MANAGEMENT

Punjab state was reorganized on 1 November 1966, and Punjab, Haryana, and Himachal Pradesh became the successor states that shared the benefits, in addition to Rajasthan. The Bhakra Management Board (BBMB) was

[2] Bhakra Beas Management Board, *History of Bhakra Nangal Project*, 1988, p. 271.

established with effect from 1 October 1967. The Board has looked after the management of the BNP since then. The development of irrigation till 1970 had to be governed in terms of the transition arrangements that are part of the Indus Treaty. The transition period ended in April 1970. Thereafter, the project development was carried out in accordance with the limitations and opportunities afforded by nature and prudent reservoir operations.

IRRIGATION DEVELOPMENT

The development of actual irrigation under the project has been summarized in Table 6.2.

It is obvious that the BNP has more than fulfilled the promised irrigation as envisaged at the time of its formulation and approval. This is due to the additional benefits reaped by the diversion of a part of the surplus Beas waters into the Bhakra reservoir since 1980 and the integrated operation of the 'eastern rivers' system.

Irrigated agriculture in Punjab and Haryana has undergone a revolution in the past three decades. Originally, the crop pattern recommended by the Crop Planning Committee was followed. The progressive farmers soon changed the crop pattern in tune with profitability and market forces. It would suffice to note that with the assistance of the government and the agricultural universities, and in tune with the demands of the new strategy on agriculture, further changes in crop pattern became necessary. These developments are discussed in the next chapter.

WATERLOGGING AND SALINITY/ALKALINITY

Irrigation is not an unmixed blessing, particularly where the farmers have a tendency to over-irrigate through excessive water applications and where improper water management practices prevail. Inadequate drainage often causes waterlogging problems, whether the lands are irrigated or not. An agricultural land is said to be waterlogged when the soil pores in the crop root zone get saturated with water. Problems of waterlogging and soil salinity/alkalinity have been universally recognized as predominantly adverse effects of unscientific irrigation. Even well-managed irrigation projects could be affected in varying degrees, depending upon the manner of water distribution, soil and crop types, and the water management practices. Waterlogging owing to periodic flooding, over-flow by run-off, seepage and impeded drainage, are common. Other activities in the project area which interfere with the drainage flow could aggravate the situation.

Table 6.2: Irrigation Development

(in thousand hectares)

Year	New Irrigation from Bhakra canals				Sirhind canal in Punjab	Total irrigation	
	Punjab	Haryana	Rajasthan	Total		Bhakra canals	Including Sirhind canal
1967–8	580	830	261	1671	905	1671	2576
1968–9	602	773	247	1622	801	1622	2523
1969–70	628	881	259	1764	906	1764	2670
1970–1	569	831	248	1648	904	1648	2552
1971–2	570	861	240	1671	903	1671	2574
1972–3	614	893	248	1755	902	1755	2657
1973–4	703	898	262	1863	896	1863	2759
1974–5	693	788	276	1757	896	1757	2653
1975–6	784	916	286	1986	896	1986	2882
1976–7	794	851	277	1922	896	1922	2818
1977–8	741	864	285	1890	990	1890	2880
1978–9	779	928	278	1985	896	1985	2881
1979–80	849	940	313	2102	1271	2102	3375
1980–1	873	1030	285	2188	1291	2188	3679
1981–2	906	1078	321	2305	1331	2305	3636
1982–3	968	1083	301	2352	1294	2352	3646
1983–4	983	1105	289	2377	1301	2377	3678
1984–5	971	969	265	2205	1301	2205	3506

Source: BBMB, *History of Bhakra-Nangal Project*, p. 280, 1988.

Waterlogging was experienced in the Indus basin in united India after irrigation was introduced on a large scale. As early as 1925, the Punjab government constituted a waterlogging enquiry committee to study and report on the extent and causes of waterlogging, which had assumed serious proportions in the irrigated areas, and to indicate preventive measures.[3] A research laboratory was established at Lahore for the analysis of soil and water and the movement of moisture in soil. In 1928, the State Waterlogging Board and Waterlogging Conference were established. Since then, considerable work on research and reclamation of affected areas was done in (United) Punjab and other provinces having similar problems.

After Independence, the Land Reclamation, Irrigation and Power Research Institute at Amritsar has continued work on these problems in India. Unfortunately, there are no accurate statistics available regarding the extent of waterlogging and the rate at which it has been increasing, except in a few states. In 1972, H.L. Uppal, former director, Land Reclamation, Irrigation and Power Research Institute, indicated that the Punjab government had launched a comprehensive programme of drainage, etc., and that a programme for tube-wells had been initiated. He further claimed 'that the high waterlogging in Punjab which posed a serious threat has been eliminated now'.[4]

The Central Board of Irrigation and Power had also reported in 1965, that in Punjab (before the bifurcation of Haryana) 0.8 million ha (2 million acres) had become waterlogged. It further noted that in 'recent years, however, there has been a significant fall in the area affected by waterlogging in the Punjab due to implementation of large scale surface drainage measures'.[5]

The Irrigation Commission Report 1972, too, indicated that the extent of waterlogging in Punjab was 'approximately 1.09 million hectares in 1958'. It added that 'since then, a number of drainage schemes have been carried out, and waterlogging has been brought under control'.[6] The specific location or canal in which it occurred was not indicated. In respect of Haryana, the Irrigation Commission reported waterlogging in 0.031 million

[3] Report of the Irrigation Commission, 1972, Government of India, Ministry of Irrigation and Power, vol. I, p. 311.

[4] Uppal, H.L., *Serious waterlogging in Punjab and Haryana—How cured and measures to prevent its recurrence*, Symposium Paper presented at 45th Annual Board Session, CBIP, December 1972.

[5] Central Board of Irrigation and Power Publication no. 76, *Development of Irrigation in India*, p. 209.

[6] Irrigation Commission Report, vol. I (see footnote above), p. 309.

ha of the Bhakra canal tract as estimated in 1965.[7] It advocated the conjunctive use of surface waters and groundwaters. It asked that farmers should be encouraged to use groundwater and noted that it helped keeping the water table down.

N.D. Gulhati, in his book on the Indus Waters Treaty, points to the US reports in the Indus valley in West Pakistan, the water logging and salinity 'situation was deteriorating so rapidly that...President Kennedy sent out a high powered Mission'. Soon after the Indus Treaty was signed, during President Ayub Khan's visit to the USA in July 1961, President Kennedy undertook to bring together a group of American experts to advise the Pakistan authorities on the problems of waterlogging and salinity. In March 1964, President Johnson transmitted to President Ayub Khan, the final report of this group.[8] The remedial efforts Pakistan made since then are beyond the scope of this study. However, an important recommendation in that report which reads as under, is equally applicable to India.

Because agriculture is beset by a wide range of impediments, the attempt to deal with any one of them in isolation will be balked by the presence of the others. Additional irrigation water, more fertilizer, improved seed and crop varieties, pest and disease control, better cultivation and salt free soil are complementary factors of production. Each may increase yields 10 to 30 per cent when applied singly, but in combination they can give increases of 200 to 300 per cent....

Report of the Working Group on Waterlogging—1992

WATERLOGGING

The Irrigation Commission estimated in 1972 that an area of 4.5 million ha had been affected all over India by waterlogging. Soon, thereafter, the Agriculture Commission estimated it in 1976 as about 6 million ha, of which 3.4 million ha was due to surface flooding and 2.6 million ha on account of a rise in the water table. The Ministry of Agriculture assumed even higher figures in 1985. The Central Water Commission gave a figure of only 1.6 million ha as affected owing to a rise in the water table.

The Coordination Committee of the Ministry of Water Resources, Department of Agriculture, and the IMD, recommended the setting-up of an

---

[7] Irrigation Commission Report, ibid., p. 310

[8] 'Report on Land and Water Development in the Indus plain'; the White House, Washington DC. US Government Printing office, 1964, p. 4, referred to by Gulhati in ref. no. 18, p. 176.

interdisciplinary working group on waterlogging, soil salinity, and alkalinity. This was set up in 1986, under the chairmanship of the adviser in the Planning Commission with members drawn from these departments. Its report published in 1992 is the most recent study on the subject.[9] As the different agencies had followed different criteria, the group standardized the norms, according to which, when the water table is within 2 metres of the land surface, the area is considered to be waterlogged. If it is 2 to 3 metres below, it is considered as a potential area for waterlogging.

The state-wise picture of waterlogging in irrigation projects in Haryana, Punjab, and Rajasthan, as per the Central Water Commission, followed by this working group, with some reconciliation of figures, is given in Table 6.3.

Table 6.3: Waterlogged areas in irrigation projects in Haryana, Punjab, and Rajasthan
(figures in 1000 ha)

State	Project	CCA	Area waterlogged
Haryana	1. Bhakra command	1166	49.17
	2 WJC command	1084	144.82
	3. Gurgaon canal	138	35.85
	Sub-total	2388	229.84
Punjab	Integrated project in south-west districts of Punjab	3072	200.00
	Sub-total	3072	200.00
Rajasthan	Indira Gandhi Nahar St. I	540	179.50
	Sub-total	540	179.50

Source: Statistical Abstracts of Haryana, Punjab and Rajasthan.

The working group reconciled estimations by different authorities at different times and updated the picture. It considered that the waterlogged area in India in the irrigation commands was 2.46 million ha. The figures reported by the Agriculture Commission and Ministry were higher because they included areas 'both due to surface flooding as well as rise in ground-water table' and were not confined to irrigated areas.

From the state-wise picture given in Table 6.3, it is relevant to note that:

(i)   The total area in India that suffers from waterlogging in irrigation commands is 2.46 million ha, half of which is in Andhra Pradesh, Bihar, and Uttar Pradesh.

[9] 'Waterlogging, Soil Salinity and Alkalinity'—Report of the working group on problem identification in irrigated areas with suggested remedial measures, December 1991, Government of India, Ministry of Water Resources, New Delhi, 1992.

(ii)   No waterlogging is reported in the Bhakra canal areas of Rajasthan.
(iii) Less than 5 per cent of the irrigated area in Haryana suffers from waterlogging. The waterlogging in Bhakra command is only 0.049 million ha, against the CCC of 1.166 million ha.
(iv)   The division of the waterlogged area in Punjab by irrigation projects has not been indicated. Hence, the extent of the area affected in Bhakra command is not given. However, the south-western districts of Faridkot, Ferozpur, and Bhatinda were reported to be affected, The working group made a specific finding that there was 'conspicuous improvement of situation in waterlogged areas of Punjab.[10] The figure of 1057,000 ha as the waterlogged area in Punjab reported by the National Commission on Agriculture (NCA) in 1976 is found to have come down to 200,000 ha This is on account of the conjunctive use of surface and groundwater, on the one hand, and the provision of a drainage component in irrigation schemes and an extensive programme of shallow tubewells on the other. The map attached to that report shows that there was no waterlogging in the Bhakra command, except perhaps for some parts of the Ghaggar bed and the area adjacent to rivers.

The report of the working group concludes with a 'Special Finding' as follows:[11]

'The outcome of the study by the Group is a conspicuous improvement of the situation in the waterlogged area in Punjab'.

SOIL SALINITY/ALKALINITY

The salinity/alkalinity in soil deposits depends on the total salt content, sodium carbonate, and bicarbonate concentration, in relation to calcium and magnesium concentration, and toxicity of specific ions of chlorides and boron. Saline soils contain excessive amounts of soluble salts. The NCA estimated in 1976 that an area of about 7 million ha in India suffers from high salinity/alkalinity, out of which 2.5 million ha is alkali soil and 4.5 million ha saline soils.

The working group classified an area as saline if the electrical conductivity (EC) was more than 4 mmhos, the exchangeable sodium percentage was less than 15 per cent and the pH less than 8.5. Areas with a pH value more than 8.5 and an exchangeable sodium percentage (ESP) more than

---

[10] Working Group Report, 1992, ibid., p. 133.
[11] Working Group Report, 1992, ibid., p. 133.

15 with EC less than 4 mmhos/cm were considered as alkali. It collected other estimates made by different agencies, reconciled the figures and considered that 3.3 million ha was affected in India by salinity and alkalinity problems. In Punjab, the extent of saline land was 0.49 million ha; Haryana had another 0.125 million ha so affected. Haryana had, in addition, 0.072 million ha of alkali land.

Himmat Singh states that 'in the post-1966 period the cultivable area in Punjab that is subject to salinity has been actually reducing steadily'.[12] According to him, surveys conducted by the Punjab Agricultural Department in coordination with remote sensing agencies has corroborated this tendency. They indicate that the cultivable area subject to salinity in Punjab state as a whole, which stood at 0.698 million ha in 1967, had come down to 0.152 million ha by 1998.

This is not to suggest that there is no room for further improvement, or to negate the need for vigilance but to drive home the fact that the picture of blighted soils and toxic environs so alarmingly painted by some 'lobby experts and obliging media' is unfounded. Generalized allegations that the 'green revolution is creating waterlogging and saline deserts in the command areas of large irrigation projects', or that huge areas of irrigated land are now waterlogged and clogged with salts, are completely unfounded.[13]

[12] Singh, Himmat, *Green Revolutions Reconsidered:The Rural World of Contemporary Punjab*, Oxford University Press, New Delhi, 2001, p. 77.

[13] Shiva, Vandana, *Staying Alive*, Kali for Women, New Delhi, 1988 p. 151, also McCully, Patrick, *Silenced Rivers: The Ecology and Politics of Large Dams*, Zed Books, London, 1996, p. 165.

7

IRRIGATED AGRICULTURE AND INCREASED PRODUCTIVITY

New Strategy for Agriculture

The approval to and the execution of the components of the BNP coincided with the early years of Independence, which brought in its trail the exodus of millions of Hindus and Sikhs from West Pakistan to India. As noted earlier, the urgent need for the rehabilitation of these migrants, as also putting back on the rails the shattered economy of East Punjab and India, played a crucial part in the planning and execution of the BNP. The most important post-Partition needs in Punjab related to productively rehabilitating the migrants, particularly experienced agriculturists, increasing production of food grains and meeting the power demands of the region. In a similar fashion, the completion of the storage dam at Bhakra leading to the development of perennial irrigation under the project, coincided with the adoption of the new strategy in agriculture, popularly known as the 'Green Revolution' in the mid-1960s in India. These opportunities were availed of by the project, to render to the nation benefits larger than originally anticipated at the stage of approval.

An effective scientific study of the effects of the BNP—in respect of agricultural production and productivity—requires complete data of the areas, production and productivity from the lands irrigated by the project over the three decades or more since the mid-sixties. In India, all agricultural and production statistics generally relate to political boundaries and administrative divisions. This has always been a handicap in scientific studies in respect of areas benefitted or affected by, say, a dam, barrage or canal. However, surrogates are often used in such cases to give a fairly accurate indication of the resulting picture. This will be the attempt in the present case too.

The lands benefitted by the irrigation canals of the BNP lie in the states of Haryana, Punjab, and Rajasthan. The irrigation command lying in Rajasthan is a small fraction of the total or cultivable area in this state, which is the largest in India area-wise. However, the project command in Haryana and Punjab covers significantly large areas of cultivated lands in the respective states. In view of these factors, and as a convenient surrogate, the agricultural production and productivity in the states of Haryana and Punjab are examined to indicate the benefits derived by these states, as also by India in general, from the BNP.

In these two states, over 80 per cent of the geographic area are sown and over 80 per cent of the net area sown is irrigated. However, in Punjab, about one-fourth of the net irrigated area is by government canals, whereas in Haryana, it is half the net area irrigated. It should be borne in mind that irrigation by tube-wells has been greatly sustained through canal water recharge, and the power supply to lift water has till recently been highly subsidized by the state through the commercial sale of BNP-generated electric energy.

Food Grains Import over the Decades

Since the Second World War, India was importing food grains to meet the minimum needs of the people. The united Punjab was one area that produced surplus food grains, which were diverted to other parts of the state and beyond. Partition robbed India of this advantage. Overnight, East Punjab needed to import food from elsewhere. India used to import, on an average, 2 to 4 million tonnes in the first decade after Independence. The increasing population pressure and periodical failure of monsoon rains brought greater shortages in their wake. The consecutive and widespread failure of rains (as in 1965 and 1966) led to greater distress. The need for larger annual imports of the order of 10 million tonnes of food grains was felt.

In the 1950s, the widespread shortage of food grains was felt in many under-developed and developing countries. The USA was then the main surplus producer and had sizeable accumulated stocks of food grains. It enacted Public Law (PL) 480 to assist poor countries, while managing to reduce its burgeoning stocks of food grains. This enabled the USA to offer many countries food grains at concessional prices and on special terms. Whether this move proved to be of help to those targetted for aid, or hindered their own efforts to achieve self-reliance, is a subject still being debated. India had then started importing wheat from the USA under PL 480. In the beginning, the imports were small, say, a million tonnes or so annually but soon they increased. Simultaneously, strings were attached and political pressures felt. There were many awkward moments in ensuring the required imports. A time came when the public distribution system was so stretched that the nation literally lived 'from ship to mouth'.

Lal Bahadur Shastri became prime minister soon after the demise of Jawaharlal Nehru in 1964. He inducted a new minister for agriculture, C. Subramaniam, who attempted to introduce many bold initiatives for sustained increase in food production. Unfortunately for them and the nation, there was a war with India's neighbour, Pakistan, and a big oil crisis in the Middle East, leading to severe pressure on the Indian economy. To make matters worse, monsoon rains failed in 1965 and again in 1966. India had to negotiate the import of 10 million tonnes from the USA that year, for 'otherwise it would have been impossible to avert human disaster of a magnitude unparalleled in human history'.

Many doomsayers predicted disaster for India. William and Paul Paddock said that there would be 'acute famine conditions and that millions and millions of people would die of starvation in the developing countries, particularly in South Asia'.[1]

The Paddock brothers classified many poor and developing nations according to their capacity for escaping famine through possible advancement in the agricultural fields. They used the principles of 'triage analysis' adopted in military medical circles which involves the classification of wounded soldiers into three major categories according to their need for urgent medical attention. They put India in the category of 'can't be saved' on the analogy of 'incurables who are generally not attended to and are left to die'. 'The hungry Nation that today refuses to heed India's history will be condemned to relive it', they concluded. Having written off India as a country for which there was no agricultural future, they further added that 'India... is the bellwether that shows the path which the others, like sheep going to the slaughter, are following. The hungry nation that today refuses to heed India's history will be condemned to relive it'.

As ill luck would have it, 1966–7 proved to be another bad year as rains failed once again. This experience was adequate warning that a country of India's size with its increasing population pressure could not be maintained on imported food grains of the order witnessed. This luckily spurred India

---

[1] Paddock, William and Paul, *Famine 1975! America's Decision: Who Will Survive?* Little Brown, Boston, 1967, quoted by Subramaniam, C. in 1979 in his lectures as Visiting Fellow at the Australian National University. Also quoted by Chopra, R.N. in *Green Revolution in India*, Intellectual Publishing House, Delhi, 1986.

to opt for a new strategy to increase its agricultural productivity and reach self-sufficiency as early as possible. 'Every country, which has improved its agriculture has done so only through the introduction of science and technology in farming. India cannot be the exception', Minister Subramaniam declared.[2] He went forward with this approach, notwithstanding resistance from various quarters. This may not be the appropriate place for a full-fledged discussion and analysis of the 'Green Revolution', or of its critics who opposed it on various premises. We shall note the consequences of the new strategy that was adopted in 1966, particularly the effects on Punjab and Haryana.

Green Revolution in Punjab and Haryana

The new strategy comprised a package of measures that included a controlled water supply through dependable irrigation, HYV seeds, fertilizer, pesticide/insecticide, and mechanical equipment. However, to optimize the results from the efforts and investments, certain preceding operations and post-harvest measures were necessary. Consolidation of fragmented land-holdings, Command Area Development (CAD) works, agricultural credit, and institutions for research, education and extension, were the important preceding steps. Rural roads, marketing, price support and remunerative price fixation were some of the other important actions required. HYV seeds set their own demanding working environment. These seeds were not high yielding by themselves but they were highly responsive to adequate inputs of irrigation and high doses of fertilizers. The exotic strains had to be adapted in India after extensive field trials. Their vulnerability to pests had to be taken care of. Full potential from the use of HYV seeds or high-responsive varieties required controlled water supply, fertilizers, pest control, agricultural research and extension and quick harvesting and land preparations, calling for some mechanization of operations.

Dr M.S. Randhawa holds that the modernization of agriculture in Punjab did not start in 1966 but really commenced much earlier with the rehabilitation of refugees in 1950 and that, by the time the HYV seeds were ready to be used in 1966, the necessary working conditions for optimized production were all broadly in place. Perhaps that is why the effects of the 'Green

[2] Subramaniam, C., *The New Strategy in Indian agriculture—the first decade and after*, Vikas Publishing House, New Delhi, 1979. Subramaniam went to Australia and spent some time as Visiting Fellow at the Australian National University, where he gave a series of seminars and lectures on India's Green Revolution. This book emanated from those lectures.

Revolution' were felt to the maximum in Punjab, Haryana and the adjoining areas of western Uttar Pradesh and Rajasthan.

The infrastructure measures adopted by the Punjab (including Haryana) government were:

- consolidation of landholdings
- encouraging and assisting conjunctive use of surface and groundwater by installation of tube-wells
- rural electrification and energization of tube-wells
- timely provision of irrigation, fertilizers/pesticides
- co-operative organizations for credit, etc.
- marketing system and villagers' link roads to markets
- quality HYV seeds
- mechanization, tractors, harvesters, etc.
- extension services and links to agricultural universities

Soon after Independence, the government of East Punjab enacted a legislation in 1948 for the consolidation of fragmented landholdings. Village advisory committees were formed for advising the staff on correcting and updating records, the classification and evaluation of fields and preparation of consolidation schemes. Irregular fields were consolidated into rectangular blocks of 1 acre (0.4 ha) each. The scheme provided for replanning the countryside including the planning of the location of schools, hospitals, and roads. Land was reserved for community centres, places of worship, and playgrounds. The village was linked with the entire cultivated area and this was, in turn, linked to the main roads. The consolidation work commenced in 1951 and was completed in the 1960s. Punjab and Haryana headed the Indian states, in the consolidation of cultivated area. Dr Randhawa holds that 'Punjab and Haryana states have the top position in the country, as the entire cultivable area in them was consolidated by 1969'.[3] This is, in fact, not a new post-Independence idea, as Punjab had the unique precedent of 'Canal Colonies' in British India in Lyallpur, Montgomery, Nilibar, etc. These canal colonies in the Indus basin brought over 10 million acres (4 million ha) of wasteland under cultivation. As the irrigation systems and canal colonies were built from scratch, due attention was paid to the civic, social, educational, and economic needs of their residents, and were based on modern town and country planning principles. They were designed to serve as trading and transport centres to serve the hinterland of the canal

---

[3] Randhawa, M.S., *A History of Agriculture in India*, Indian Council of Agricultural Research, New Delhi, 1986, vol. IV, Chapter 3, p. 36.

irrigated colony. The effort was not merely restricted to agricultural development but social engineering in a rural context.[4] These villages really constituted new trailblazing settlements.

While the construction of the Nangal barrage and canal system by 1954 enabled kharif irrigation, the completion of the Bhakra dam and the conclusion of the Indus Treaty cleared the way for assured perennial irrigation for substantial areas. Thus, by the time a new thrust was given to agriculture by the introduction of HYV seeds and the related package of measures, Punjab and Haryana were ready to achieve optimum benefits thereof. Even then, however, there were many critics of and political controversies regarding the new strategy. Some sociologists too voiced apprehensions.

What were those apprehensions like? C. Subramaniam, Minister of Agriculture, states some of the objections that he faced and his reasoning thereof. The traditionalists thought that India had a well-established tradition in agriculture and that there was nothing that our farmers needed to learn from others. The scientific approach to agriculture was a fad that would lead to disasters. Subramaniam thought that it would not be derogatory to the prestige of our ancestors, or to our present-day farmers, if we discarded outdated ideas and outmoded tools in agriculture. If we had to increase our yields to a level not previously conceived, we had to alter the whole set of agricultural practices that were suited for an entirely different production range and food needs.

Some sociologists felt that it would be the big farmers who would reap the advantage, while small farmers would be left behind. As a result, disparities between big and small farmers would widen. The minister said that India had hardly any other option in the light of her food needs and the already very high imports. He was in favour of high production and achieving self-sufficiency within India, and facing the social tensions by distributing food grains on an appropriate welfare basis.

A paper prepared by B.S. Minhas and T.N. Srinivasan created another controversy. This purported to prove statistically that if the available fertilizers were distributed at the rate of only 10 kg/ha or less, the total yield of food grains would be higher than if we used the same total quantity in a concentrated way in specific areas.[5] The statistical analysis might have been correct but what they had failed to note was that there was a difference in the fertilizer response from traditional and HYV seeds. In low doses of

[4] Malcolm, L. Darling, *Punjab Peasant in Prosperity and Debt*, Oxford University Press, London, 1947.

[5] Minhas, B.S. and Srinivasan, T.N., 'Agricultural Strategy Analysis', *Yojana*, vol. 10, no. 1, 1966, p. 21.

up to about 10 kg/ha or so, there was imperceptible yield gain in HYV seeds; thereafter, the yields dramatically shot up. At higher dosages, production for HYV seeds was 200 per cent or more.

Another controversy related to the use of chemical fertilizers. Opponents of chemical fertilizers claimed that it burned up the soil and rendered it useless, despite the fact that other countries had used it for half a century. The government was not against the use of organic manure but the difficulty was the high quantum of such needs and the managerial aspects of applying them at all the critical phases of growth. This controversy still continues. It is, however, true that because of intensive cultivation, micronutrients need to be replenished.

Some politically aligned groups, particularly the communists, took the view that all this was an American idea, and it was their way of establishing domination over India in agriculture. That it should, therefore, be resisted. Indira Gandhi, who had meanwhile become prime minister, supported her agriculture minister. A determined government pushed through the Green Revolution and soon the results were there for all to see. Now that some thirty-five years have elapsed, it would be worthwhile to take stock of these measures.

Dr Norman Borlaug, who later became director of the wheat department of the International Maize and Wheat Improvement Centre (CIMMYT) in Mexico, produced HYV seeds capable of responding to very high doses of fertilizers and irrigation. He visited India in 1963, and on his return, sent a 100 kg of seed of each of the dwarf and semi-dwarf wheat varieties. These were tested at multiple locations in India. From them, two varieties, which ultimately proved popular, were Kalyan Sona and Sonalika. During 1966–7, India imported 18,000 tonnes of HYV wheat seeds, Lerma Roja 64, and Sonora 64, with the assistance of the Rockefeller Foundation. Never before in the history of agriculture had the transportation of a large quantity of HYV seeds coupled with an entirely new technology and strategy, been achieved on such a massive scale in so short a time, and with such great success, observed Dr Borlaug.

A beginning was made, as seen above, with wheat. Similarly, the International Rice Research Institute (IRRI) worked on rice seeds. IR-8, the first of the new varieties produced in 1966 at IRRI, gave thrice the yield of the conventional varieties. Initially, the emphasis was on increasing the yield and improving grain quality. Later, incorporation of insect and disease resistance received priority. Further improved varieties followed. India cooperated with the IRRI and was one of the first countries to cultivate IR-8 on a massive scale. The Central Rice Research Institute, Cuttack, undertook breeding programmes on rice seeds. In 1968, the variety Jaya was released. Many other varieties followed in the early 1970s. Rice is grown during the monsoon season when there is plenty of moisture present and the conditions favour pests and diseases. Moreover, the new varieties required controlled irrigation but our traditional canal systems did not provide for such control of each field. Nevertheless, rice production rose but perhaps not as dramatically as in the case of wheat.

Even by the year 1970, Punjab and Haryana were intensely cultivated. These two were the only states of India that had over 80 per cent of the total area as the net sown area. There was not much additional area available for cultivation. Hence, any increase in the crop area could only be achieved through multiple cropping which could be done with assured irrigation. This was, in fact, what these states did.

The highest intensity of cropping for any state of India is found in Punjab, which also has the highest development of irrigation in India. B.D. Dhawan, who made a study of the trends and new tendencies in Indian irrigated agriculture in 1993, based on the data for the time series for 1950–1 to 1987–8, came to the same conclusion with respect to Punjab. He also pointed out that even in the drought year 1987–8, the state recorded a cropping intensity of about 176 per cent, on a net sown area of the order of 4 million ha., which has perceptibly risen at the annual rate of 0.27 per cent between 1967–8 and 1987–8. The intensity of cropping is quite high in Haryana, too, but it varied over time. Dhawan had pointed out that 'though it rose between 1967 and 1987 at the annual rate of 0.62 per cent, the rise was accompanied by a Coefficient of Variation (CV) around a trend line of 4.3 per cent, well above the corresponding CV of 1.1 per cent for Punjab'. In the drought year 1987–8, the crop intensity was 145.1 per cent. A clearer picture is given in Table 7.1.

The Punjab farmers were indeed the pioneers of the new technology. They set the pace for the agricultural revolution in India. The most important wheat growing areas of India were Punjab, Haryana, and adjoining western Uttar Pradesh. The results were soon visible to all. Punjab, which was producing less than 2 million tonnes of wheat in 1965–6, stepped it up to over 5 million tonnes by 1970–1, crossed 6 million tonnes by 1976–7, and reached 10 million tonnes by 1984–5. By 1990–1, Punjab produced 12 million tonnes of wheat and went beyond 15 million tonnes by the year 2000.

If the Green Revolution started with wheat, it was soon followed by the rice revolution. Prior to 1965–6, rice was grown only in some pockets in the state where there was clayey soil and assured irrigation facilities. Now,

Table 7.1: Progressive Land Use in Punjab and Haryana

(All figures of areas are in 1000 ha)

Year	Net sown area	Col. 2 as % of total area	Total cropped area	Col. 4 as % of col. 2	Net irrigated area	Col. 6 as % of col. 2	Gross irrigated area	Col.8 as % of col. 2
1	2	3	4	5	6	7	8	9
Punjab—Total area 5035,000 hectares								
1966–7	3757	75	4732	126	2020	54		
1970–1	4053	81	5678	140	2888	71	4245	105
1980–1	4191	83	6763	161	3382	81	5781	138
1990–1	4218	84	7502	178	3909	93	7055	167
1999–2000	4237	84	7847	185	3977	94	7544	178
Haryana—Total area 4421,000 hectares								
1966–7	3423	77	4599	134	1293	38	1736	51
1970–1	3565	81	4957	139	1532	43	2230	63
1980–1	3602	81	5462	152	2134	59	3309	92
1990–1	3575	81	5919	166	2600	73	4237	119
1999–2000	3552	80	6029	170	2888	81	5124	144

Source: Statistical Abstracts of Punjab and Haryana—various years.

with the extension of irrigation and the advent of HYV rice seeds, rice was cultivated in areas which could not have supported its cultivation earlier. From a production of less than half a million tonnes of rice in 1968–9, it crossed 3 million tonnes by 1978–9 and touched 5 million tonnes by 1985. In the year 2000, it stood at 9 million tonnes. Haryana, which emerged as a new state in 1966, was not far behind Punjab from which it was carved out, though it could not boast of as extensive an irrigation system as Punjab. The wheat production in Haryana was just a million tonnes in 1966–7. It crossed 3 million tonnes by 1978–9 and reached 5 million tonnes by 1985. Rice production by 1999–2000 in Haryana was well over 9 million tonnes.

Details of the annual area, production, and yield of wheat and rice crops in Punjab and Haryana have been presented in Tables 7.2 to 7.5.

Impact of the Green Revolution

There have been many studies on the various impacts of the 'Green Revolution' in India. Most studies have acknowledged the significant step-up to India's food production that it enabled and its other positive impacts. However, some of these studies have ended with criticism. Sizeable literature, which examines the 'Green Revolution' from various viewpoints, has built up during the last three decades. Selective quotations from them have often been used, either to praise or denigrate the 'Green Revolution' and its impact. A discussion thereof is outside the scope of the present study. However, the salient criticisms made could be taken note of in order to see if these would be applicable in the context of the BNP, or in the wider sense in respect of the states of Punjab and Haryana.

The generally voiced criticism on the impact of 'Green Revolution' (setting aside some of the polemical, extreme views which have no inherent merit), are as under:

(i)   The 'Green Revolution' was, at best, a wheat revolution, which had no effect on the other crops.

(ii)   Agricultural growth and increased productivity by large inputs of chemical fertilizers were not sustained, and they disappeared after the initial years. The chemical fertilizers led to environmental instability.

(iii)   There was stagnation in food production.

(iv)   It led to regional and social imbalances. The 'Green Revolution' was accompanied by inequity. Farmers with marginal and small holdings benefitted much less than those with large holdings.

Table 7.2: Area, Production, and Yield of Wheat in Punjab

Year	Area (in 1000 ha)	Production (in 1000 tonnes)	Yield in Kg/ha	% increase decrease in yield	HYV Area (in 1000 ha)	% area under HYV seeds	Irrigated wheat area in 1000 ha (% of wheat area)
1965–6	1548	1916	1104	–	–	–	–
1966–7	1608	2449	1520	+37.7	–	–	–
1967–8	1790	3335	1863	+22.6	621	34.7	–
1968–9	2063	4491	2177	+16.85	–	–	–
1969–70	2191	4918	2245	+3.1	–	–	–
1970–1	2299	5145	2238	–0.3	1589	69.1	1942 (84.5%)
1971–2	2336	5618	2406	+7.5	–	–	–
1972–3	2404	5368	2233	–7.2	–	–	–
1973–4	2338	5181	2216	–0.8	–	–	–
1974–5	2213	5300	2395	+8.1	–	–	–
1975–6	2439	5788	2373	–0.9	–	–	–
1976–7	2579	6272	2432	+2.5	–	–	–
1977–8	2617	6642	2538	+4.4	–	–	–
1978–9	2736	7423	2713	+6.9	–	–	–
1979–80	2823	7896	2797	+3.1	–	–	–
1980–1	2812	7677	2730	–2.4	2757	98	2567 (91.3%)
1981–2	2917	8553	2932	+7.4	–	–	–
1982–3	3054	9183	3007	+2.6	–	–	–
1983–4	3125	9422	3015	+0.3	–	–	–
1984–5	3095	10,176	3288	+9.1	–	–	–
1985–6	3112	10,988	3531	+7.4	–	–	–
1986–7	3189	9458	2966	–16.0	–	–	–
1987–8	3139	11,084	3531	+19.0	–	–	–
1988–9	3156	11,580	3669	+3.9	–	–	–
1989–90	3251	11,681	3593	–2.1	–	–	–
1990–1	3272	12,155	3715	+3.4	3271	99.9	3144 (96%)
1991–2	3233	12,295	3803	+2.4	–	–	–
1992–3	3281	12,369	3770	–0.9	–	–	–
1993–4	3326	13,341	4011	+6.4	–	–	–
1994–5	3311	13,542	4090	+2.0	–	–	–
1995–6	3221	12,510	3883	–5.1	–	–	–
1996–7	3229	13,672	4234	+9.0	3229	100	–
1997–8	3301	12,752	3853	–8.3	3301	100	–
1998–9	3278	14,192	4332	+12.4	3278	100	–
1999–2000	3388	15,910	4696	+8.4	3388	100	–

Source: *Statistical Abstract of Punjab—various years.*

Table 7.3: Area, Production, and Yield of Rice in Punjab

Year	Area (in 1000 ha)	Production (in 1000 tonnes)	Yield in Kg/ha	% increase decrease in yield	HYV Area (in 1000 ha)	% area under HYV seeds	Irrigated rice area in 1000 ha (% of rice area)
1966–7	285	338	1186	–	–	–	–
1967–8	–	–	–	–	–	–	–
1968–9	338	460	1361	–	–	–	–
1969–70	384	573	1492	+ 9.6	–	–	–
1970–1	390	688	1765	+18.3	130	33.3	358 (91.8)
1971–2	450	920	2044	+15.8	–	–	–
1972–3	476	955	2008	–1.8	–	–	–
1973–4	520	1189	2287	+12.2	–	–	–
1974–5	569	1179	2072	–9.4	–	–	–
1975–6	567	1447	2552	+23.2	–	–	–
1976–7	674	1741	2583	+1.2	–	–	–
1977–8	831	2494	3001	+16.2	–	–	–
1978–9	1052	3091	2938	–2.1	–	–	–
1979–80	1167	3041	2606	–11.3	–	–	–
1980–1	1183	3233	2733	+4.9	1095	92.6	1157 (97.8)
1981–2	1270	3755	2957	+8.2	–	–	–
1982–3	1319	4147	3144	+6.3	–	–	–
1983–4	1481	4536	3063	–2.6	–	–	–
1984–5	1644	5052	3073	+0.5	–	–	–
1985–6	1714	5449	3179	+3.4	–	–	–
1986–7	1809	6022	3329	+4.7	–	–	–
1987–8	1720	5442	3164	–5.0	–	–	–
1988–9	1778	4925	2770	–12.5	–	–	–
1989–90	1908	6697	3510	+26.7	–	–	–
1990–1	2015	6506	3229	–8.0	1906	94.6	1998 (99.1)
1991–2	2074	6755	3257	+0.9	–	–	2054
1992–3	2065	7002	3391	+4.1	–	–	2058 (99.3)
1993–4	2174	7624	3507	+3.4	–	–	2162
1994–5	2265	7703	3382	–3.6	–	–	2228
1995–6	2185	6843	3131	–7.4	–	–	2175 (99.4)
1996–7	2159	7334	3397	+8.5	–	94.5	2137 (99.0)
1997–8	2278	7890	3465	+2.0	–	94.5	2262
1998–9	2518	7993	3152	–9.0	–	95.1	2448
1999–2000	2604	8716	3347	+6.2	2604	–	2585 (99.3)

Source: *Statistical Abstract of Punjab—various years.*

Table 7.4: Area, Production, and Yield of Wheat in Haryana

Year	Area (in 1000 ha)	Production (in 1000 tonnes)	Yield in Kg/ha	% increase decrease in yield	HYV Area (in 1000 ha)	% area under HYV seeds	Irrigated wheat area in 1000 ha (% of wheat area)
1966–7	743	1059	1425	–	–	–	–
1967–8	–	–	–	–	–	–	–
1968–9	895	1522	1700	–	–	–	–
1969–70	1017	2120	2085	+22.7	–	–	–
1970–1	1129	2342	2074	–0.5	630	55.8	914 (81.0%)
1971–2	1177	2402	2041	–1.6	–	–	–
1972–3	1270	2231	1757	–13.9	–	–	–
1973–4	1178	1811	1537	–12.5	–	–	–
1974–5	1117	1954	1749	+12.1	–	–	–
1975–6	1226	2428	1980	+13.2	1087	88.7	1084 (88.4%)
1976–7	1348	2735	2029	+2.5	–	–	–
1977–8	1360	2845	2092	+3.1	–	–	–
1978–9	1481	3398	2294	+9.7	–	–	–
1979–80	1471	3283	2232	–2.7	–	–	–
1980–1	1480	3492	2359	+5 .7	1360	92.0	1378 (93.1%)
1981–2	1562	3682	2357	–0.1	–	–	–
1982–3	1720	4345	2526	+7.2	–	–	–
1983–4	1784	4458	2499	–1.1	–	–	–
1984–5	1705	4421	2593	+3.6	–	–	–
1985–6	1701	5260	3094	+19.3	1612	94.8	1623 (95.4%)
1986–7	1782	5055	2836	–8.3	–	–	–
1987–8	1731	4861	2808	–1.0	–	–	–
1988–9	1827	6255	3424	+21.9	–	–	–
1989–90	1859	5913	3181	–7.1	–	–	–
1990–1	1850	6436	3479	+9.4	1829	98.9	1805 (97.6%)
1991–2	1808	6502	3596	+3.4	–	–	–
1992–3	1956	7083	3621	+0.7	–	–	–
1993–4	1998	7231	3619	–0.1	–	–	–
1994–5	1986	7303	3677	+1.6	–	–	–
1995–6	1972	7291	3697	+0.5	1863	94.5	1939 (98.3%)
1996–7	2017	7826	3880	+4.9	–	–	–
1997–8	2064	7554	3660	–5.7	–	–	–
1998–9	2188	8568	3916	+7.0	2135	97.6	2148 (98.5%)
1999–2000	2316	9650	4166	+6.4	2260	97.6	2288 (98.8%)

Source: *Statistical Abstract of Haryana—various years.*

Table 7.5: Area, Production, and Yield of Rice in Haryana

Year	Area (in 1000 ha)	Production (in 1000 tonnes)	Yield in Kg/ha	% increase decrease in yield	HYV Area (in 1000 ha)	% area under HYV seeds	Irrigated rice area in 1000 ha (% of rice area)
1966–7	192	223	1161	–	Nil	Nil	138 (71.9%)
1967–8	–	–	–	–	–	–	–
1968–9	223	265	1190	–	–	–	–
1969–70	241	371	1540	–	–	–	–
1970–1	269	460	1710	+11.0	30	11.1	235 (87.4%)
1971–2	291	536	1842	+7.7	–	–	–
1972–3	291	462	1588	–13.8	–	–	–
1973–4	292	540	1851	+16.6	–	–	–
1974–5	275	393	1426	–23.0	–	–	–
1975–6	303	625	2063	+44.7	–	–	–
1976–7	330	815	2470	+19.7	–	–	–
1977–8	372	965	2601	+5.3	–	–	–
1978–9	459	1228	2678	+3.0	–	–	–
1979–80	509	942	1851	–30.9	–	–	–
1980–1	484	1259	2602	+40.6	414	85.6	469 (96.9%)
1981–2	506	1250	2470	–5.1	–	–	–
1982–3	489	1275	2607	+5.5	–	–	–
1983–4	533	1325	2486	–4.6	–	–	–
1984–5	557	1363	2447	–1.6	–	–	–
1985–6	584	1633	2796	+14.3	–	–	–
1986–7	628	1543	2457	–12.1	–	–	–
1987–8	462	1073	2322	–5.1	–	–	–
1988–9	599	1437	2399	+3.3	–	–	–
1989–90	621	1698	2734	+14.0	–	–	–
1990–1	661	1834	2775	+1.5	479	72.4	655 (99%)
1991–2	640	1812	2831	+2.0	–	–	–
1992–3	703	1869	2659	–6.1	–	–	–
1993–4	753	2057	2732	+2.9	–	–	–
1994–5	795	2227	2802	+2.6	–	–	–
1995–6	830	1847	2225	–20.6	498	60.0	824 (99.3%)
1996–7	831	2463	2964	+33.2	–	–	–
1997–8	913	2556	2800	–5.5	–	–	–
1998–9	1086	2432	2239	–20.0	674	62.1	1084 (99.8%)
1999–2000	1083	2583	2385	+6.5	620	57.2	1081 (99.8%)

Source: *Statistical Abstract of Haryana—various years.*

(v) It led to farm modernization and seasonal unemployment at the national level.

Over three decades of the actual impact experienced due to the introduction of large-scale irrigation in Punjab and Haryana and the adoption of new strategies and package of measures popularly known as the 'Green Revolution' are now available before the nation. This affords an opportunity to examine whether these various allegations were warranted, in this regard.

A careful examination of these allegations with respect to their possible applicability to the present case leads to the conclusion that these are inapplicable as well as extremely exaggerated as general points. This position is briefly discussed in the following paragraphs.

It was not Merely a Wheat Revolution

It is incorrect to state that increased production or improved productivity was limited to wheat in Punjab and Haryana. The improvement was experienced in all the major crops grown.

The growth rate of food grain production between 1960–1 and 1978–9, on an all-India basis, was 2.77 per cent, but varied from state to state. Punjab and Haryana led with average growth rates of 8 per cent and 5.3 per cent per year respectively.[6] Jawaharlal Nehru University (JNU) and the Planning Commission jointly made an analysis of growth at the district level. Forty-eight districts recorded a high growth rate of over 4.0 per cent annually during 1962–65 to 1971–74. Punjab, Rajasthan, Haryana, and western Uttar Pradesh accounted for three-fourths of these districts.[7]

On an all-India level, the term 'food grains' does not necessarily limit itself to cereals, or rice and wheat. However, with respect to Punjab and Haryana, particularly after the BNP and the adoption of the tube-well programme, rice and wheat were the overwhelmingly important food crops grown. This has continued over the last few decades.

The progressive increased production in food grains is summarized in Table 7.6.

Table 7.6 clearly shows that the production increase was noted both in Punjab and Haryana with respect to all food crops grown in these states, which includes not only wheat but also rice. The trend of increase in production has continued till date.

[6] Sarma, J.S., *Agricultural Policy in India: Growth with Equity*, International Development Research Centre, Canada, 1982, p. 27.

[7] Attributed to G.S. Bhalla and Y.K. Alagh, *Spatial Pattern of Levels and Growth of Agricultural Output in India*, Jawaharlal Nehru University, New Delhi (Mimeo).

Table 7.6: Progressive production of food grains

(million tonnes)

Year	India			Punjab			Haryana		
	Food grains	Rice	Wheat	Food grains	Rice	Wheat	Food grains	Rice	Wheat
1966–7	72.4	30.4	11.4	4.17	0.34	2.45	2.59	0.22	1.06
1970–1	108.4	42.2	23.8	7.31	0.69	5.14	4.77	0.46	2.34
1980–1	129.6	53.6	36.3	11.90	3.23	7.68	6.04	1.26	3.49
1990–1	176.4	74.3	55.1	19.25	6.51	12.16	9.56	1.83	6.44
1999–2000	209.8	89.7	76.4	25.20	8.72	15.91	13.07	2.58	9.65

Source: Statistical Abstracts of India, Punjab and Haryana—various years.

INCREASED PRODUCTIVITY WAS SUSTAINED

As indicated in Table 7.1, in both Punjab and Haryana the net sown area has remained more or less the same from 1970–1 till the present day. The increased crop area under rice and wheat was due to multiple cropping on the same net area, enabled by irrigation. The increased production over the last three decades is therefore, more attributable to improved productivity. The yield of wheat and rice from year to year has been presented in Tables 7.2 to 7.5.

Table 7.7 gives a summary of the progressive improvement in the productivity of rice and wheat in Haryana and Punjab. It clearly shows that the yields of rice and wheat have been progressively rising in Punjab and Haryana for the last three decades and more. Moreover, the level of productivity achieved in these states was always significantly higher than the all-India average. This trend of increased productivity is still continuing. The Planning Commission's Tenth Five Year Plan document confirms that the annual compound rates of growth (percentage) of crop yield (at a constant price of 1990–3) for 1980–95 is higher than the corresponding figures for 1970–83, in both Haryana and Punjab.[8]

To ensure sustained environmental stability and soil health, organic compost, bio-fertilizers and micronutrients should form an integral part of balanced fertilizer application and integrated nutrient management. Increased production and marketing of quality bio-fertilizers should be promoted.

THERE WAS NO STAGNATION IN FOOD PRODUCTION

Those who speak of stagnation or slow rate of growth in production of food grains in independent India would do well to note that according to S.R. Sen in the entire first half of the twentieth century, that is, between 1901 and 1950 food grain production in India was stagnant.[9] For undivided India, the food grain production in 1947 was 66 million tonnes, which was less than the production in 1900, which was 67 million tonnes. George Blyn's study of agricultural output in India from 1891 to 1947 confirms Sen's results.[10]

[8] The Planning Commission, *Tenth Five Year Plan* (2002–7), New Delhi, 2002, vol. III, p. 78.

[9] Sen, S.R., *Growth and Instability in Indian Agriculture, Agricultural Situation in India*, vol. XXI, no. 1, January 1967, Ministry of Food and Agriculture, Government of India.

[10] Blyn, George, *Agricultural Trends in India from 1891 to 1947*, University of Pennsylvania Press, 1966.

Table 7.7: Progressive improvement in yield of major food crops

(kilograms per hectare)

Year	Food grains			Rice			Wheat		
	India	Punjab	Haryana	India	Punjab	Haryana	India	Punjab	Haryana
1966–7	644	1259	736	863	1186	1161	887	1520	1425
1970–1	872	1861	1235	1123	1765	1710	1307	2238	2074
1980–1	1023	2458	1518	1336	2733	2602	1630	2730	2359
1990–1	1380	3390	2351	1740	3227	2775	2281	3715	3479
1999–2000	1704	4028	3047	1994	3347	2385	2778	4696	4166

Source: Statistical Abstracts of Punjab and Haryana.

The better performance of independent India in terms of food production during the next 50 years (1951–2000), particularly since the 'Green Revolution' (1970) clearly stands out. This achievement is however, sought to be underrated owing to the population growth in the corresponding period, which has shot up well beyond the projections of the planners.

REGIONAL IMBALANCE

It is true that there has been differential improvement in the production and productivity levels in different states in India. The bulk of the increase in the output of food grains is concentrated in a few regions. There has, thus, been inter-regional inequality in growth. The Government of India's Seventh Five Year Plan document admitted in 1985 that the 'differential pattern and pace of development, particularly the growth of food grains production, has led to regional disparities.'[11]

In India, the climatic conditions, soils, extent of irrigation coverage, degree of application of various inputs such as fertilizers, the position of land reforms achieved, etc., are all different in the various states. The prices of inputs and outputs have also varied over different periods. All these aspects need to be analysed to explain the causes for regional imbalances. It would be, in any case, unrealistic to expect uniform growth of food grain production/productivity in all the varying agro-climatic regions of the country. It may be more sensible to help each region make optimum use of its development potential, instead of seeking uniformity in the growth production process.

Even within the same state, some districts have recorded higher yields than others. For instance, even within the progressive Punjab state, such differences are noted. In 1999–2000, the rice yield was 3611 kg/ha in Ludhiana, while it was 2831 kg/ha in Gurdaspur. Similarly the wheat yield in that year was 5064 kg/ha in Ludhiana and 3591 kg/ha in Hoshiarpur.

C.H. Hanumantha Rao pointed out in 1994 that:

An encouraging feature of agricultural growth during the eighties is its steadiness and its spread to the regions and crops which were lagging behind in the first decade of the 'Green Revolution'.[12] The performance of the eastern and western regions, which are relatively more dependent on the monsoons than the northern region, has shown an improvement in the recent period.

[11] Government of India, *Seventh Five Year Plan (1985–90)*, Planning Commission, Delhi, vol. I, p. 14, 1985.

[12] Rao, Hanumantha, C.H., *Agricultural Growth, Rural Poverty and Environmental Degradation in India*, Oxford University Press, New Delhi, 1994, p. 244.

The Tenth Five Year Plan document (2003) notes that during the period 1980–3 and 1992–5, a 'very high rate of growth of productivity of 3.15 per cent per annum was recorded. This growth was more or less equally experienced by all the regions'.[13]

For the purpose of this study, it is enough to reiterate that the north-western region of India, particularly Punjab and Haryana, have gained significantly through the 'Green Revolution'.

SOCIAL INTER-CLASS DISPARITIES

Once the innovation of HYV seeds and irrigation broke the technological barrier to reward human endeavour, the responsibility was cast on economic and institutional policies (and policy makers) to ensure that the technological gains were not allowed to be monopolized only by the privileged and the powerful. It is possible to raise the income from crop production of small and marginal farmers by providing them preferred access to credit, inputs and marketing, and extension services. It may perhaps be conceded that in the first decade of the 'Green Revolution', there were inegalitarian effects and an aggravation of inter-class disparities. However, it must be recognized that many studies of the distribution of gains of technological changes are vitiated by the 'single factor' analysis. While analysing additional gains to different classes of producers, one should not merely look for changes in production. Changes in prices and changes in the shares of different classes of agriculturists in the total area cultivated are also very relevant for such studies.

A common complaint made is that increasing dependence on purchased inputs had driven small farmers with limited access to resources and credit out of cultivation. However, M.L. Dantwala points out that in Punjab, small farmers do not seem to have had much difficulty in obtaining purchased inputs.[14] He cited Bhalla and Chadha, who showed that differences by size-class in total bio-chemical inputs per acre are not significant. As a result, for Punjab as a whole, farm business income per acre does not show any significant size-class difference. The position of Haryana is similar.

The important question that we face is really whether the new agricultural strategy was responsible for the proletarianization of Indian agriculture, if it existed in any region.

---

[13] The Planning Commission, *Tenth Five Year Plan (2002–7)*, New Delhi, 2002, Volume III, p. 78.

[14] Dantwala, M.L., *Dilemmas of Growth*, Sage Publications, New Delhi, 1996, p. 137.

The basic Green Revolution Technology (GRT) model is on the premise that such technology is scale-neutral and equally indifferent to the amount of capital employed. In other words, the size of a holding is not related to the efficient utilization of the GRT. It does not preclude small holdings from its adoption. Thus, small landholders should gain equally. The experience of Punjab seems to vindicate this premise. The 'Green Revolution' has not aggravated poverty or inequality in Punjab but has increased the incomes of both small and large farmers. There are many studies which conclude that small farmers and agricultural labour have, in fact, benefitted more from the 'Green Revolution' than large land owners.

Studies conducted by the Programme Evaluation Organization, Planning Commission, did not show any bias in terms of farm size, with regard to the adoption of HYV of wheat in Punjab, although small farmers were a little slower in adopting the new varieties.[15]

A field study done for Punjab in 1974–5 by Bhalla and Chadha shows a revealing picture of the impact of the 'Green Revolution' on input and productivity in relation to farm size.[16] This study found that there is no significant difference between different-sized groups of farms with regard to the percentage of the net operated area irrigated. Despite this, the cropping intensity is significantly higher among smaller farms due to the larger availability of family labour per acre. The slight edge that large farmers had with regard to inputs was more than counter-balanced by the small farmers' advantage with regard to availability of labour.

Farm Mechanization did not Lead to Seasonal Unemployment

Farm mechanization by using tractors, tillers, disc-harrows, pumps, etc., is important for intensive irrigated cultivation. If the rice-wheat cycle of cropping is practised, the farmer has hardly two to three weeks' time to complete the harvesting and clear the stubble of the earlier crop, to be ready for the next one.

The general allegation is that the adoption of farm mechanization induced by the 'Green Revolution' displaces labour and creates unemployment.

Punjab and Haryana makes the most extensive use of labour-saving with machinery like tractors. There is, however, no evidence to show that labour

[15] Sarma, J.S., *Agricultural Policy in India: Growth with Equity*, International Development Research Centre, Canada, 1982, p. 40.

[16] Bhalla, G.S. and Chadha, G.K., *The Green Revolution and the Small Peasant*, Concept Publishing Company, New Delhi, 1983.

input has decreased because of this. On the other hand, available evidence shows that labour use and wages have improved since the 'Green Revolution' was adopted. The Birla Institute of Scientific Research (BISR) made a study of agricultural growth and employment shifts in Punjab.[17] The study says that 'the intensity of human labour used per unit area was 66 per cent higher in 1976–7 than in 1967–8'. In Punjab, the share of workers in the secondary and tertiary sectors, which was on a decline till 1971, also increased thereafter.

FEMALE DISADVANTAGE AND THE GREEN REVOLUTION

In recent decades, much literature has emerged on the relationship between social and economic development and female disadvantage, particularly in the Asian and Indian context. Worldwide efforts to remove gender-based inequalities received momentum during the UN Women's Decade (1975–85). Similarly, in the south Asian region, the status of the girl child received focus when 1990 was observed as the SAARC Year of the Girl Child. The report of the 'Committee on the Status of Women in India' titled *Towards Equality* was one of the earliest in India, which highlighted the fact that despite the progressive provisions of the Indian Constitution, development had been accompanied by a deterioration in the situation of women.

The Indian Constitution accords the highest importance to the welfare and advancement of women and children. It has an impressive list of provisions seeking to remove every kind of discrimination against women. It even empowers the state to make affirmative discrimination in favour of women and children. It enjoins the state to provide equal and adequate means of livelihood, and equal pay for equal work. Numerous legislative measures too have been taken to give practical shape to the constitutional directives.

A recent study released in October 2003 by the Registrar General of India and the Census Commissioner, the Ministry of Health and Family Welfare and the UN Population Fund, brings out the reality that modern India still clings to its old beliefs and attitudes. The sex ratio in India has been unfavourable to women and girls and has been deteriorating from decade to decade. In the age group 0–6 years, it has decreased at a much faster rate than the overall ratio. In addition, the position in some of the states is worse than the average picture for India as seen in Table 7.8.

[17] BISR, *Agricultural Growth and Employment Shifts in Punjab*, New Delhi, 1981.

Table 7.8: Sex ratios in some states of India, 2001

State	Sex ratio	Sex ratio (0–6 years)
India	933	927
Delhi	821	865
Haryana	861	820
Himachal Pradesh	970	897
Punjab	874	793
Rajasthan	922	909

Source: Census of India.

The comment of the Union Health Minister, Sushma Swaraj, on this situation is very eloquent. She said, 'if the enactment of a law was the only thing needed to curb this menace, this would have stopped long ago'... 'But it is the social belief that having sons is the only way to continue the family lineage that perpetuates this', she added.

The Department of Women and Child Development in India concedes that 'discrimination in the nutritional habits among boys and girls is evident from the fact that the nutritional intake for girls is inferior in quality and quantity; boys have access to higher value foods; boys are given the first priority for the available food within the family... the root cause is not so much the lack of food but the lack of value attached to the girl child'.[18]

Many indicators have been used by different commentators to substantiate the bias against women. These include the worsening sex ratio, declining female work participation, persistent shortfall in nutrition, low literacy and high mortality. A gendered understanding of development or poverty also raises questions relating to the reliability of the indicators that highlight gender disadvantages. Besides, can these indicators lend themselves to different and contrary meanings if they are uprooted from their context, thereby leading some 'committed analysts' to opt for a meaning favouring their views? A discussion on these issues is beyond the scope of the present study. However, note has to be made of some criticism which uses the gender angle to deride the quest for greater agricultural production, as also the Green Revolution.

S. Sudha and S. Irudaya Rajan dealt with female disadvantage, particularly in infancy and childhood, owing to long-standing social patterns of

[18] Government of India, Ministry of Human Resource Development, Department of Women and Child Development, Annual Report 2002–3, pp. 56–7.

preference for male children.[19] They suggested that parents, particularly in north and north-west India and many urban areas, may be adding pre-natal sex selection techniques to traditional post-natal ones to create a 'double jeopardy' for their daughters. They alleged that the evidence from Punjab, Haryana, Uttar Pradesh, and Tamil Nadu shows that the Green Revolution adversely affected women's work participation.

Dr Vandana Shiva, noted environmentalist, considers that marginalization of women is an inevitable result of most development programmes based on scientific and economic paradigms created by western gender-based ideology.[20] She believes that the Green Revolution is a western 'masculinist paradigm' of food production, which has destroyed women's traditional role in food production, agriculture and forestry. According to Shiva, the new technology excludes women and marginalizes them. This devaluation, combined with increased work burdens, reduces women's entitlement to food, nutrition and even life itself. Thus, women are increasingly seen as a burden on society and can be dispensed with through discrimination, death and foeticide. Many do not agree with her generalizations and somewhat extreme views, but an examination of the demerits thereof is outside the scope of this study. However, the specific allegations levelled by her about the Green Revolution in north-west India could be summed up as under:

(i)   It concentrates on the already privileged farmers.

(ii) Organic manure is no longer the fertilizer; it is fertilizer factories that are the only sources of soil fertility.

(iii) 'In the heartland of the Green Revolution region of Punjab, the food abundance for the market has not been translated into nutrition for the girl child within the house'. A study done in 1978 in the Ludhiana district of Punjab showed that the 'percentage of female children who were undernourished was higher than that of undernourished male children within the same economic group'.

(iv) 'Violence against women related to dowry issues has been found to be the highest in the Green Revolution region of north-west India and is part of the general violence that is becoming endemic to Punjab.' The Green Revolution as the breeding ground for civil

---

[19] S., Sudha and S. Irudaya Rajan, *Female Demographic Disadvantage in India, 1981–1991, Sex Selective Abortions and Female Infanticide*, Working Paper No. 288 of the Centre for Development Studies, Thiruvananthapuram, 1997 (revised version included in Shahra Razavi, *Gendered Poverty and Well Being*, Blackwell Publishers (for Institute of Social Studies), 2000, p. 176.

[20] Shiva, Vandana, *Staying Alive, Women, Ecology and Development in India*, Kali for Women, New Delhi, 1988, Chapter 5.

unrest and violence in the state has been analysed by Shiva in a study carried out for the UN University in 1987.[21]

(v) Shiva quotes approvingly from Bina Agarwal who has noted that it is the north-western region where discrimination against females is most noted, both historically and in the recent period.[22]

Gender disparities exist everywhere in the world. There would be general agreement that we must fight against gender inequity, exploitation, and repression but most would not agree with Shiva equating women and nature, as if *only women and all women* are conservationists and equity-oriented. It is not women alone who are involved in the struggles for the protection of the rights of minorities and marginal sections. Her exaggerated claim that the development and prosperity, caused by the Green Revolution or otherwise, is the cause of violence against women, or violence in general, is unacceptable. There are numerous nations of the world that are less affluent but, nonetheless, continue to exhibit gender-balanced demographic measures. The proposition that economic development devalues women is not enough reason for families to discriminate against their daughters. Social and cultural factors are equally responsible for all variations in the status of women. Shahra Razavi of the UN Research Institute for Social Development is of the view that the evidence on female disadvantage in early age survivorship does not 'imply a consistent pattern of anti-female bias in food intake and nutritional status as is often assumed.'[23] She specifically points out that 'even in north and north-western India, where the evidence on discrimination against young girls in terms of survivorship is most compelling, findings from nutritional surveys not infrequently shows that adult women fare better than their male counterparts.'

Confirming the problematic nature of these generalized assumptions on female disadvantage, R. Saith and B. Harris-White's comprehensive review concludes that the evidence on gender differentials in nutritional status is inconclusive, showing no consistent indication of a gender bias.[24] They also point out that 'significant proportions of the conditions that cause morbidity are sex-specific and defy simple male/female comparisons.' Somewhat similar questions can be raised about life expectancy data, too. The point

---

[21] Shiva, Vandana, *Violence and Natural Resource Conflict: A Case Study of Punjab*, Report for UNU, Tokyo, 1987.

[22] Agarwal, Bina, op. cit.

[23] Razavi, Shahra, *Gendered Poverty and Well Being* © Institute of Social Studies, The Hague, Blackwell Publishers, Oxford, UK, 2000.

[24] R., Saith and Harris-White, B., see Razavi, Shahra (2000), ibid., pp. 57–77.

of raising these issues here is to highlight the methodological controversies and arbitrariness involved in making comparisons between men and women as regards their well-being.

Conclusion

Food grain production in India has generally kept ahead of the population growth rate, throughout the period since the 1950s. There were, however, sub-periods of much greater growth or a relatively slower rate. Till 1965 or so, growth in area contributed to increased production, though significant gains in yields were recorded. The increase in production since around 1971 is more due to increased productivity. Gains in food grains production have been predominantly in wheat and rice. The main source of increase in the productivity of wheat and rice since the mid-1960s has been the introduction of irrigation and the adoption of HYV of seeds. HYV wheat seeds were progressively introduced from 1967, while the introduction of HYV rice seeds was in the early 1970s.

The significantly high productivity achieved in Punjab and Haryana under the 'Green Revolution' has been attributed to several factors. These include, in addition to the general factors mentioned earlier, assured irrigation, land reforms like consolidation of the farm holdings prior to the introduction of irrigation, the onset of the Green Revolution and the peasant-proprietor land tenure system. Other factors include successive, farmer-friendly, vigilant and responsive governments, an enterprising farming community, groundwater development and incentives in inputs and procurement.

The salient effects thereof, *ex post facto*, with respect to Punjab and Haryana (the latter being an integral part of the former till 1966) based on recorded data, can now be summed up.

The picture obtained is as follows:

1.   There has been a continuing upward trend in the production and productivity of rice and wheat, both in Punjab and Haryana. The yield in these states now is about three times that of 1967. In the case of rice, it is three and a half times that of 1967 in Punjab. However, in Haryana, it is around two and a half times that of 1967. It is worth noting that in both the states, almost the entire rice and wheat crop is irrigated. Average rice yields have fluctuated from year to year with variation in the quantum and distribution of rainfall. Moreover, HYV seeds cover only 60 per cent of the Haryana rice crop so far, as against 95 per cent in Punjab.

2. The average wheat yield in Punjab is the highest for any state in India. Haryana comes second. Only these two states in India have crossed yields of over 4 tonnes per hectare. The average yield of rice is also the highest in Punjab. The traditional rice-growing states of Tamil Nadu and Andhra Pradesh follow close behind.

3. Continuing improvements in the yield of wheat and rice have shown that the exaggerated apprehensions held in the past by critics of the Green Revolution were not justified, at least in this region.

4. There is enough empirical evidence to show that irrigation and an improved package of measures reduce instability in production and yield, benefiting thereby both the farmers and the national economy.

5. The people of Punjab and Haryana are not traditional rice growers or consumers. Yet, the farmers here are apparently willing to take chances for maximizing their gains. The large-scale production of rice in this region since the 1970s is a result thereof. This has proved to be a major blessing not merely for them but also for the people depending on the public distribution system in India.

6. Punjab and Haryana contributed half to three-fourths of the wheat and rice that came to the central pool in the past three decades.

7. Another positive outcome of the rural prosperity triggered by the Green Revolution is the near absence of rural out-migration since 1966 in Punjab and Haryana.

8. The percentage contribution to the GDP from agriculture and allied sectors in Punjab and Haryana has remained steady in the last three decades. Significant industrial growth also occurred in these states during this period.

9. The states have effectively tackled the groundwater rise noticed in the initial period after the introduction of large-scale irrigation. The encouragement given by the government for conjunctive use of groundwater and surface water by offering incentives for the development of tube-wells has paid off. The high water table and the tendency to waterlog remains confined to a small region around Faridkot and Ferozpur in Punjab. Besides, the cultivable area subject to salinity in Punjab has been continuously reducing.

10. A general condemnation of the instrumentalities of irrigation, or of the Green Revolution is therefore, unwarranted. The application of these measures in the ambience and regulated environment of Punjab (including Haryana) has yielded significant social and economic benefits without experiencing any unacceptable environmental costs.

8

PRODUCTION OF FOOD GRAINS

Contribution of Large Dams to India's Food Production

'Contribution of large dams poor', screamed the headline in the widely respected Indian newspaper in September 2000.[1] The point was further amplified in the body of the story thus:

The contribution of large dams to increased food grains production in India is only 10 per cent contrary to largely held belief, is a finding of the India Country Study (ICS) on large dams conducted by a consultant team of prominent Indian experts for the World Commission on Dams (WCD).

Was that correct reporting? Was that ICS report based on reliable data and adequate analyses? These were the natural reactions that arose in the minds of the discerning public, particularly the post-Independence generation, who were assured that these large dams were the harbingers of prosperity. Another important question that arose was whether these dams would have any useful role to perform in the future, for instance, by way of enabling food self-sufficiency and power generation in India.

First, the basic facts. Apart from the hundreds of representations and views received by it, the WCD itself commissioned the preparation of many working papers.[2] All of them, whether duly checked and verified or otherwise, formed part of its 'knowledge base'. In fact, a criticism of the WCD procedure was that it did not verify or check the correctness of the information presented to it by various interested parties. The WCD selected India

[1] See the *Hindu*, New Delhi, 24 September 2000. Report by Gargi Parsai. Similar news reports had appeared in the *Tribune* of 23 September and the *Statesman* of the 24 September. Perhaps all were inspired by the same source.

[2] The WCD knowledge base consists of seven case studies, two country studies, one briefing paper, seventeen thematic reviews of five sectors, a cross-check survey of 125 dams, four regional consultations and nearly 950 topic-related submissions.

for a case study in the context of its global review of the development effectiveness of large dams and their consequences. The Indian team presented its report to the WCD in June 2000.The above-mentioned news item was supposedly based on this report.[3]

The preface of the ICS itself points out that 'it became clear at an early stage that for various reasons this could not be a joint report of the group as a whole, but would have to be a collection of papers by the different members on various aspects'. The responsibility for each of the chapters lay with the respective authors and the others were not necessarily in agreement with all that was said in the papers other than their own. However, all of them came together in the writing of the last chapter—some agreed conclusions—by consensus. This important point must be kept in view when one considers any point made anywhere in that report.

Analysis of the Controversy

The statement referred to earlier about the contribution of large dams was obviously based on para 2.4 of Chapter II, authored by Nirmal Sengupta of the Madras Institute of Development Studies (MIDS). According to him, there were many claimants to the credit for the achievement of self-sufficiency in food grains production, including irrigation, measures for productivity increase through HYV of seeds, increased fertilizer use, and promotion of agricultural research/extension/credit. He thereafter tried to apportion the credit between them.

The purported view of the team was, in fact, the view of one member, with which the others were not in agreement. Chapter I of the ICS had already indicated that the World Bank Performance Review (1998) showed 'that irrigation has played a core role in agricultural production and growth' (see Para 1.3). The report pointed out that 'irrigation enables a higher productive potential from the land and significant production response from associated use of high yielding varieties, chemical fertilizers and other inputs. After pointing out the difficulty in ascertaining the precise contribution of irrigation, it concluded that 'nevertheless, various estimates point to a contribution from irrigated-agriculture to overall agricultural production of about two-thirds and under some circumstances an even higher contribution'.

[3] Rangachari, R., *et al.*, June 2000 (mimeo), *Large Dams: India's Experience A Report Prepared for the World Commission on Dams*. The five-member team comprised R. Rangachari, Nirmal Sengupta, Ramaswamy R. Iyer, Pranab Banerji, and Shekhar Singh.

The consensus of all the members, reported in Chapter 7, was as follows:

The production of food grains increased from 51 million tonnes in 1950–51 to almost 200 million tonnes in 1996–97. About two thirds (66.7%) of this increase came from the irrigated area, which is around one-third of the cultivated area. The increase in production of food grains was the result of a combination of several factors, such as HYV seeds, chemical fertilizers...clearly irrigation played a crucial role, and some of that irrigation came from large dams.

As regards the quantum of increase attributed to dams, it mentions the different views held—one putting it at 10 per cent and others who considered 'that this is an underestimate, as irrigation is a precondition for the use of the other inputs' (see para 7.1—subpara 7). They concluded that large dams have made a contribution to the development of irrigation and, therefore, to food production and food security. The final 'summing up' (See para 7.6) was that 'large dams have made important contributions to the development of irrigated-agriculture and improved productivity and the production of food'.

Input-output relations in agriculture are complex, non-linear and interactive. Ignoring this is a basic flaw in the study by Nirmal Sengupta. He further ignored the synergy between irrigation and fertilizers, under which a higher level of irrigation was associated with higher efficiency of fertilizer use. A. Vaidyanathan, also of MIDS, had pointed out that 'attempts at applying multiple regression analysis to assess the relative contributions of rainfall, irrigation, irrigation quality, and fertilizers in explaining spatial and temporal variations in productivity have not been successful or conclusive'.[4] He had further shown that in a given rainfall zone, districts with higher levels of irrigation used more fertilizers per hectare and produced higher value of output per hectare.

The progressive growth in area, production and yield in respect of all food grains in India during the last fifty years is summarized in Table 8.1.

Sengupta compared the food grain output of 1950–1 with that of 1993–4. He apportioned the increase to an increase in gross area under food grains and improved productivity and attributed only the balance to irrigation. This was irrational and unjustified. The lack of requisite data for such a study, as also the drawbacks in his approach have been pointed out.

In addition, his study ignored an important feature in the expansion of cropped area. For some years after independence, an increase was

[4] Vaidyanathan, A., *Water Resource Management: Institutions and Irrigation Development in India*, Oxford University Press, New Delhi, 1999, p. 79.

Table 8.1: Growth of Area, Production and Yield in food grains

Year	Area (million ha)	Production (million tonnes)	Yield (kg/ha)
1950–1	97.3	50.8	522
1955–6	110.6	66.8	605
1960–1	115.6	76.7	662
1965–6	115.1	72.3	629
1970–1	124.3	108.4	872
1975–6	128.2	121.0	944
1980–1	126.7	129.6	1023
1985–6	128.0	150.4	1175
1990–1	127.8	176.4	1380
1993–4	122.8	184.3	1501
1995–6	121.0 ·	180.4	1491
1999–2000	123.1	209.8	1704

Source: *Agricultural Statistics of India, Ministry of Agriculture and CMIE, Mumbai, Economic Intelligence Service.*

reported in the net cultivated area. The net sown area in 1950–1 was 118.8 million ha, which increased to 140.3 million ha by 1970–1. Thereafter, it generally hovered within a very narrow range between 140 and 142.5 million ha. In 1993–4, it was even slightly lower than what it was twenty years earlier in 1973–4. However, the *gross cropped* area showed a rising trend, increasing from 131.9 M. ha in 1950–1 to 186.4 in 1993–4, mostly enabled by an increase in the *irrigated area.*

In the same manner, the area under all food grains that stood at around 100 million ha in the early 1950s, crossed 124 million hectares in 1970–1. Throughout the 1990s, it was around 123–4 million hectares only. However, production and productivity were generally rising.

In the allocation of credit attempted by Sengupta, the period from 1950–1 and 1970–1 should have been taken as one block and between 1970–1 and 1993–4 as the second block. The picture would then be modified as shown in Table 8.2.

Thus, the conclusion drawn by Sengupta that the contribution of large dams to increased food grains production was marginal and less than 10 per cent was incorrect and unsubstantiated. Another of his conclusions that 'rain-fed farming had responded well to productivity increase measures', was once again, incorrect and unsubstantiated. Unless the total rainfall in any selected year was adequate for the full crop needs and its distribution

Table 8.2: Area under food grains and Production (1950–1 to 1993–4)

	Unirrigated land	Irrigated land	Total	Observed
Gross area under food grains—1950–1 (in million ha)	79.0	18.3	97.3	97.3
Gross area under food grains—1970–1 (in million ha)	94.2	30.1	124.3	124.3
Gross area under food grain—1993–4 (in million ha)	76.45	48.25	124.7	124.7
Productivity 1950–1 (t/ha)*	0.4	1.0	Average	0.522@
Productivity 1970–1 (t/ha)	0.7	1.4	Average	0.872#
Productivity 1993–4 (t/ha)*	1.0	2.3	Average	1.501$
Production 1950–1 (million tonnes) (1)×(4)	31.60	18.30	49.90	50.8
Production 1970–1 (million tonnes) (2)×(5)	65.94	42.14	108.08	108.4
Production 1993–4 (million tonnes) (3)×(6)	76.45	110.98	187.43	185.0
Increase in production between 1950–1 and 1970–1 in million tonnes	34.34	23.84	58.18	–
Increase in production between 1970–1 and 1993–4 in million tonnes	10.51	68.84	79.35	–
Increased production between blocks of years—in percentage of the total				
1950–1 to 1970–1	59%	41%	100%	–
1970–1 to 1993–4	13.25%	86.75%	100%	–

Source: CWC and Nirmal Sengupta.
Notes: *assumed by Sengupta.*
@ and # taken from *Agricultural Statistics at a Glance*, Ministry of Agriculture, 1992.
$ taken from *Water and Related Statistics*, Central Water Commission, 1998 which, in turn, has attributed it to the Ministry of Agriculture.

over the crop period had synchronized with the crop requirements, the requisite moisture regime cannot be provided and the productivity and crop yields would be subnormal, if anything at all. The farmers were unlikely to gamble with inputs if water was not assured. The fact that the productivity averaged 1 t/ha in the early 1990s and was *not* expected to cross 1.5 t/ha *even* after about sixty years is eloquent. On the other hand, the irrigated crop was projected to reach the average productivity level of 4 t/ha Besides, the net cropped area in India cannot increase over the present levels, except in a very marginal manner. Hence, future food production would have to largely depend on irrigated agriculture.

The *Economic Survey* (2001–2) indicates the annual compound growth rate of crop area, production and productivity of food grains in India, as shown in Table 8.3.[5] It is seen that the production of rice and wheat, the principal cereals consumed in India, recorded in the last two decades an annual growth rate greater than the growth rate of population (1.9 per cent). However, pulses and coarse cereals have not maintained an adequate growth rate in the last decade on an all-India basis.

Table 8.3: Annual compound growth of area, production, and productivity

(%)

Crop	1980–1 to 1989–90			1990–1 to 2000–1		
	Area	Production	Yield	Area	Production	Yield
Rice	0.41	3.62	3.19	0.63	1.79	1.16
Wheat	0.47	3.57	3.10	1.21	3.04	1.81
Rice & wheat	0.43	3.59	3.15	0.84	2.27	1.42
Food grains	–0.23	2.85	2.74	–0.20	1.66	1.34

Source: *Government of India, Economic Survey—2001–02.*

The Institute of Economic Growth published a study by B.D. Dhawan on *Trends and New Tendencies in Irrigated Agriculture.*[6] His book presents some interesting results of his studies on yields from irrigated and unirrigated lands. He has asserted, based on his analysis, that 'irrigated-agriculture alone provides the hope for the needed production of food grains through

[5] The Government of India, *Economic Survey*, 2001–2.

[6] Dhawan, B.D., *Trends and Tendencies in Indian Irrigated Agriculture*, Commonwealth Publishers, New Delhi for the Institute of Economic Growth, Delhi, 1993.

a continued step up in crop yields along a steady path, besides the expansion in intensity of cropping'. There are other relevant findings, too. The study in respect of different states, selected and analysed by him, showed that overall irrigated yields were more than twice the corresponding unirrigated yield levels. There is an absence of any upward trend in yields in unirrigated lands. In contrast, the overall yield of the irrigated segment has consistently risen at 2–3 per cent per annum.

In his concluding remarks (pages 78–9), he recommends that 'we should persist with our past policy of irrigation being the kingpin of our agricultural development. Otherwise, a major policy change shift in gear in favour of dryland farming, currently emanating from anti-big dam forces and dryland farming proponents can prove very costly from the point of view of self sufficiency in food grains in the next two decades'.

The results of a study on the relative productivity of irrigated and unirrigated lands in various states of India have been reported in a publication by A.Vaidyanathan of MIDS, the same institution in which Nirmal Sengupta worked.[7] The relevant portion reads as follows:

Trend analysis of (time series estimates of the productivity of irrigated and rain-fed crops covering the 1970s and 1980s) indicates that, in a large majority of States, irrigated yields show a statistically significant rising trend over this period even as production per unit of unirrigated area does not exhibit a statistically significant trend in either direction. Productivity of unirrigated land is thus more or less stagnant and the productivity differential between irrigated and rain-fed areas has progressively widened both in absolute and relative terms. Moreover, irrigated yields are generally more stable than rain-fed yields.

It is surprising that Sengupta did not take note of studies made in his own organization.

Dams by themselves do not lead to increased agricultural production, unless there is an efficient irrigation canal network, and the irrigated agriculture is backed by a well-planned package, in an ambience of agrarian and related reforms. Even HYV seeds by themselves cannot give high yields. They are highly responsive to ideal inputs of irrigation and fertilizers. Sengupta lays all the alleged adverse effects of major and medium irrigation projects at the doors of large dams. However, their beneficial effects are sought to be whittled down on the plea that not all major and medium irrigation projects are dam-backed. Further, he attributes to 'large dams' all ills arising out of drift in governmental policies in regard to irrigation

[7] Ibid.

projects, lack of political will to levy and collect appropriate irrigation water rates, indifference to the enforcement of agrarian reforms and lack of proper pricing policy on food output or fertilizer input.

National Commission for Integrated Water Resources Development

The NCIWRD was set up by the Government of India to prepare an integrated water plan for the development of India's water resources. The commission was set up in September 1996 and its report was submitted to the Government of India in September 1999.

It was presided over by Dr S.R. Hashim, Member, Planning Commission and had a number of distinguished members.

The commission made projections of the population likely over the next few decades, as also their long-term food demand estimates. Using the high and low population projections and per capita per annum food requirements, they made projections of food production needs for the year 2010, 2025, and 2050. For meeting these domestic demands for food grains, the water demands for irrigation for the various years were also worked out based on many assumptions.[8]

The projections are summarized in Table 8.4.

Table 8.4: Projections of the National Commission

	Year 2010	Year 2025	Year 2050
Net cultivable area M.ha	143.0	144.0	145.0
Total cropped area M.ha	193.0	202 to 204	218 to 232
Productivity—unirrigated land T/ha	1.1	1.25	1.5
Productivity—irrigated land T/ha	3.0	3.4	4.0
Food grain needs—low demand M.tonnes	245	308	420
Food grain needs—high demand M.tonnes	247	320	494

Source: *NCIWRD Report, Volume I, Table 3.25, p. 59.*

The NCIWRD has reposed faith in irrigated agriculture as the means to meet India's food demands. Its projections on meeting the requirements in the next five decades are summarized in Table 8.5.

[8] Ministry of Water Resources, September 1999, Report of the National Commission for Integrated Water Resources Development (NCIWRD) vol. I, p. 57.

Table 8.5: Projections by the NCIWRD Plan

	Unirrigated crop	Irrigated crop	Total
Food Grains Production (FGP) 1993–4 M.t.	76.45	110.98	187.43*
FGP projection 2010 low demand (L.D.) M.t.	84.0	162.3	246.3
FGP projection 2010 high demand (H.D.) M.t.	83.0	166.2	249.2
FGP projection 2025—low demand M.t.	91.5	215.5	307.0
FGP projection 2025—high demand M.t	87.75	233.56	321.31
FGP projection 2050—low demand M.t	103.35	316.8	420.15
FGP projection 2050—high demand M.t	85	409.0	494.0
FGP increase 1993–4 to 2010 low demand M.t.	7.55	51.32	58.87
FGP increase 1993–4 to 2010 high demand M.t.	6.55	55.22	61.77
FGP increase 2010–25 low demand M.t.	7.5	53.2	60.7
FGP increase 2010–25 high demand M.t.	4.75	67.36	72.11
FGP increase 2025–50 low demand M.t.	11.85	101.3	113.15
FGP increase 2025–50 high demand M.t.	-2.75	175.44	172.69
FGP increase needed in blocks of years expressed as percentage of the total			
Between 1993–4 and 2010 low demand	12.8%	87.2%	100%
Between 1993–4 and 2010 high demand	10.6%	89.4%	100%
Between 2010 and 2025 low demand	12.4%	87.6%	100%
Between 2010 and 2025 high demand	6.6%	93.4%	100%
Between 2025 and 2050 low demand	10.5%	89.5%	100%
Between 2025 and 2050 high demand	-1.6%	101.6%	100%

Source: Report of the NCIWRD and the discussion materials circulated to its members.

Note: * The earlier projection was 187.43 MT. However the actual production recorded was 184.26 MT. only. The projected figure has been retained here for uniformity.

Conclusions

The findings and conclusions are summarized below:

1. It is incorrect to state that the five-member team that prepared the India Country Study (June 2000) found the contribution of large dams to increased food grains production poor, or less than 10 per cent. This was the view of one member which was based on inferences and arguments. The other members of the ICS team did not endorse it.

2. Such 'arithmetical allocations of contributions' overlook the fact that a number of factors are involved in increasing agricultural productivity and production like irrigation, improved package of inputs such as fertilizers, HYV seeds, and credit. However *irrigation is a precondition without which nothing would have reliably worked*. It will not be possible to isolate these factors and the interaction and synergy among them, and attribute percentage contributions with clinical precision.

3. Irrigation and large dams have indeed played a very significant role in increased food production and greater self-reliance. They will have to play a continued role to fulfil the nation's needs in the next fifty years.

4. The NCIWRD has clearly underlined the greater and overwhelming role of irrigated agriculture for meeting India's food needs during the next five decades, during which the population is projected to rise before it stabilizes.

Contributions of Punjab and Haryana

The basic salient details of land use can be summarized as in Table 8.6.

The states of Punjab and Haryana as they are today, constituted one province prior to 1966. The UBD Canal system, the Sirhind canal system, the BNP, and Western Jamuna canal are the major irrigation systems of Punjab. If we are to ascertain the role of the irrigation-backed Green Revolution in the food production of India, there can be no better case study than this one.

Food Grains

Punjab and Haryana together account for more than a third of the total wheat production in India. No other Indian state has yet reached the consistently

Table 8.6: Land use in Punjab and Haryana (2000)

S. no.	Item	Unit	Haryana	Punjab
1.	Geographic area	Million ha	4.402	5.036
2.	Net area sown	Million ha	3.526	4.237
3.	Total cropped area	Million ha	6.115	7.847
4.	Area under rice	Million ha	1.083	2.604
5.	Area under wheat	Million ha	2.354	3.388
6.	Net area irrigated	Million ha	2.888	3.977
7.	Gross area irrigated	Million ha	5.223	7.544
8.	Net area irrigated as % of net sown area	%	81	94
9.	Gross irrigated area as % of total cultivated area.	%	85	96
10.	Net area irrigated by canals	Million ha	1.441	0.977
11.	Net area irrigated by tubewells	Million ha	1.432	2.982

Source: Statistical Abstracts of Punjab (2000) and Haryana (2001).

high productivity levels of these states. The highest productivity in respect of the rice crop in India is also that of Punjab (over 3.5 tonnes/ha), followed by the traditional rice growing states of Tamil Nadu, Andhra Pradesh, Karnataka, and West Bengal.

Punjab and Haryana not only have high production levels of wheat and rice but also contribute the maximum to India's Public Distribution System (PDS), which seeks to enhance food security, especially for the economically weaker sections of society. During the period 1996 to 2000, Punjab's contribution alone to the central rice procurement was around 40 per cent of the total. The corresponding wheat contribution was around 60 per cent. Throughout the 1980s, the average contribution of Punjab and Haryana has been around half in respect of rice and two-thirds in case of wheat. For decades, they have remained, very important for their contribution to the central procurement pool of food grains for public distribution. This will be seen from Table 8.7.

The Government of India's Economic Survey (2001–2) points out that 'it is clear that only three States, Punjab, Haryana, and Andhra Pradesh contribute to the bulk of the procurement'. In 2001–2, Punjab and Haryana alone accounted for over 80 per cent of the total wheat procurement. Similarly, in that year, Punjab contributed over a third of the procurement of rice. The 2002–3 *Economic Survey* says that 'Andhra Pradesh, Punjab,

Table 8.7: Procurement of food grains in the 1980s (according to crop year)

(in million tonnes)

State/region/year	1980–1	1981–2	1982–3	1983–4	1984–5	1985–6	1986–7	1987–8	1988–9	1989–90
RICE
All India	5.61	7.33	7.05	7.73	9.86	9.88	9.16	6.90	7.73	11.86
Haryana	0.67	0.88	0.71	0.62	0.98	1.03	0.68	0.32	0.67	0.96
Punjab	2.53	3.11	3.25	3.27	4.29	4.22	4.38	3.37	2.86	5.00
WHEAT
All India	6.60	7.72	8.29	9.30	10.35	10.54	7.88	6.54	9.00	11.07
Haryana	1.12	1.26	1.40	1.77	1.96	2.34	2.25	1.26	1.97	2.59
Punjab	3.77	4.83	5.18	5.01	6.15	6.48	4.42	4.75	5.60	6.74

Source: Directorate of Economics and Statistics, Ministry of Agriculture, Government of India, Agricultural Statistics at a Glance, 1992, p. 75.

Haryana are the major rice procuring states, while Punjab, Haryana and Uttar Pradesh account for the bulk of wheat procurement'.[9] It will be no exaggeration to say that these two states have been the kingpins of public distribution in India and that the BNP has played a very crucial role in this.[10]

The Centre for Monitoring the Indian Economy closely monitors the Indian economy and reports on it. Its *Economic Intelligence Service on Agriculture* indicates the top ten districts in the production of principal agricultural commodities. According to its report for the triennium 1991–3, the top ten districts in wheat production were Sangrur, Ferozpur, Faridkot, Amritsar, Patiala, Ludhiana, Bhatinda, and Jalandhar in Punjab and Hisar in Haryana, as well as Ganganagar in Rajasthan.[11] Similarly, Sangrur and Patiala are among the top ten districts in rice production too. Ganganagar tops the list in production of gram. Bhatinda, Faridkot, Ganganagar, Hisar, Ferozpur, and Sirsa are among the top ten districts in respect of cotton production. These districts were the top ten in the previous triennium 1988–91 too. All these districts are the beneficiaries of the Bhakra dam. Of course, these districts have developed their groundwater too. However, the significant impact of the BNP in these record productions cannot be disputed.

---

[9] Government of India, *Economic Survey*, 2002–3, p. 88.

[10] A common criticism of the PDS is that while the government has burgeoning stocks of food grains, millions of poor go to bed hungry. The government needs to take such appropriate policy decisions that obviate this paradox of poverty amidst plenty. The PDS has proved to be a great boon—and in states where it is well run, it enables millions of poor people to eat better. In Kerala, for instance, severe malnutrition among children was less than 5 per cent compared to 18 per cent nationally.

[11] Centre for Monitoring the Indian Economy, Mumbai, *Economic Intelligence Service-Agriculture*, December, 2002.

9

HYDROELECTRIC POWER

Early Proposals

When the Bhakra dam proposal was first mooted in 1910, the purpose of the scheme was mainly to provide irrigation in the arid and semi-arid tracts of land between the Sutlej and the Yamuna. Hydroelectric power generation was not envisaged as one of its objectives. Even the 1919 Project Report only highlighted the devastation caused by famine in the areas that it proposed to cover by irrigation. It did not include hydropower as an objective. The 1939–42 Project Report incorporated hydroelectric power development, for the first time, as an integral part of the project. It provided for four units of 40 MW each, with a fifth unit of the same size as an auxiliary unit. In the 1945–6 Project Report and the designs prepared by International Engineering Company (IECO), USA, hydropower was highlighted and the firm power indicated as 150 MW. The 1949 Project Report, prepared after Indian Independence, also provided for four units at Bhakra in the first stage.

The revised Project Report of 1951 was a multi-purpose scheme, incorporating irrigation and hydropower as the primary objectives. However, the August 1951 Report did not include provisions for installation of power units as the Central Electricity Commission had made a very low load forecast in the region. The emphasis had shifted by then to the execution of the Nangal barrage and canal system, as also the two powerhouses on the NHC. In 1952, the Punjab Electricity Department reviewed the load forecast and recommended the concurrent installation of two units at Bhakra, and the postponement of the installation of the third unit at the two powerhouses on the NHC. However, the installation of only one unit at Bhakra and the completion of civil works for another unit as also the service bay, were agreed upon. The position soon changed. In March 1955, the BCB decided on the concurrent installation of four units at Bhakra. By 1960, the forecast indicated a further need, for the fifth unit at Bhakra, as well as the third unit in the two powerhouses on the NHC. The sixth and seventh units at Bhakra were indicated as required by, 1962 and 1965.

Load Estimation and Phased Power Production

Planning for power had been one of the major controversial issues in several projects undertaken in those days. Bhakra was no exception to this. The project, envisaged in the early fifties, provided for ten power generation units of 100 MW each in the powerhouses at Bhakra and six units of 24 MW each in the powerhouses on the NHC. The Bhakra dam power plant I would be on the left side of the river, with its centre line 500 feet downstream of and parallel to the axis of the dam. It was to have six units ultimately. Power plant II, on the right side of the river, would be similarly located, with its centre line at about 660 feet from the axis of the dam and would have four units. However, this plant was to be installed at a later date, as and when the load demanded.

The initial installation at Bhakra was to comprise a full structure for four units alone, and a service bay with one installed unit.

The philosophy of the Planning Commission has been that additions to capital equipment should be made in a progressive and systematic manner, so as to derive the maximum benefits from the investments made. The First Five Year Plan had recognized that 'cheap electric power is essential for the development of the country. In fact, modern life depended so largely on the use of electricity that the quantity of electricity used per capita in a country is an index of its material development and of the standard of living attained in it'.[1] It can not only improve agricultural production and encourage cottage and small-scale industries but also make life in rural India attractive, thus arresting the urban influx. Unfortunately, some planners and economists were of the firm conviction that it would be unsound planning to have the Bhakra-Nangal power plants produce as much as 400 MW of power. At that early stage of India's development, with hardly 14 kWh being the annual per capita power use, they conceded in principle, that we had a long way to go. Yet, when it came to project approval, plan allocation, and budget provisions, they showed their 'go slow' powers.

The First Plan considered the anticipated load by 1956 as 69 MW, and it provided for an installed capacity of 96 MW under the BNP. It, however, qualified this by stating that the load figure was tentative, pending the results

---

[1] The Planning Commission, *First Five Year Plan, 1952*, p. 345.

of load surveys. The Second Five Year Plan had to admit that, 'during the last few years, the rate of utilization of power has been more rapid than before. It is likely to rise. The present estimates of loads for the next ten years will probably have to be revised upwards.' The Third Five Year plan again confessed that 'detailed load surveys conducted by Central Water and Power Commission towards the end of the Second plan indicated much higher demands than were assessed', in 1958.

In retrospect, the planners were repeatedly found to have been less than farsighted. Consequently, there was a delay in approving the power component of the project as originally envisaged. Till mid-1954, the view held was that all requisite power could be generated by the Nangal hydel powerhouses. The planners considered that the need for the first unit of 90 MW at the Bhakra dam would not arise till 1969. Even when work was started on the dam, it was believed that the energy generated at the left bank power plants would not be fully used for a number of years, and hence, the second powerhouse on the right bank would not be needed for a long time. The reality turned out to be very different. The rapid demand for power even before the completion of the power plants on the left bank necessitated concurrent planning for the right bank power plants. Eventually, by 1969, all the ten units at the Bhakra dam were functional, in addition to the six units at the two powerhouses on the NHC; yet there remained some unsatisfied demands.

Perhaps wisened by the events, Jawaharlal Nehru, who headed the Planning Commission, had a better vision of the likely developing power demands in his later days. He said at the Conference of the Ministers of Irrigation and Power on 3 January 1964, 'as for electric power the more I think of it the more I feel the importance of it. It does not matter how much electric power you have in India, it will always fall short of the demand'.

It is interesting to note that even four decades after Jawaharlal Nehru made that statement, the US Department of Energy stated in April 2004 that 'developing world's demand for energy is projected to roughly double over the next two decades as economies and population in developing countries grow faster than those in industrialised countries'. The International Energy Outlook 2004 shows the strongest growth in energy consumption among the developing nations of the world, especially developing Asia (including China and India). Another highlight of that report is about the electric power use. It says that 'the world net electricity consumption nearly doubles over the projection period, from 13,290 billion kWh in 2001 to 23,072 billion kWh in 2025. Strong growth in electricity use, averaging 3.5 per cent per year is projected for the developing world, where robust economic expansion drives demand for electricity....'[2]

Powerhouses on NHC

The NHC begins immediately upstream of the Nangal barrage on the left bank of the Sutlej. It is a fully lined channel, 61.06 km. (37.94 miles) long. The channel runs along the foot of the Siwalik hills through an extremely difficult sub-mountainous tract divided by hill torrents. The natural fall in the ground level along the channel is utilized at Ganguwal and Kotla for generating hydropower. The formation of the channel involved deep cutting in hillocks and heavy filling in the ravines. There are more than fifty-five torrents along the channel, which it crosses through siphons, aqueducts or superpassages, as was considered appropriate at the time. The NHC has a designed carrying capacity of 12,500 cusec (354 cumec). A silt ejector has been provided in the head reach and, for this purpose, an additional capacity of 2000 cusec is provided till the ejector. The normal slope given is 1 in 10,000. However, steeper slopes are given in deep cuttings. The normal section involves a bed width of 92.6 ft (28.2 m), depth of 20.6 ft (6.3 m) and side slopes of 1.25: 1.

Nangal powerhouse 1 is located at Ganguwal, about 12 miles (19 km) from the Nangal barrage. Powerhouse 2 is located at Kotla, about 6 miles (10 km) downstream. An average head of about 93.5 ft (28.35 m) is available at each of these powerhouses. Originally, these two constituted the grid, and the load dispatch centre was set up at Ganguwal. After the construction of the Bhakra dam and the commissioning of the left bank powerhouse, the power stations at Ganguwal and Kotla became virtual base load stations.

Bhakra Dam Powerhouses

There are two power plants at the Bhakra dam, with five generating units in each. Each unit has a separate penstock These welded plate steel penstocks, 4.57 m (15 ft) diameter each, have been installed, five each on either side of the spillway in the non-overflow section of the dam, for supplying water for power generation. Each group passing through the powerhouse is located on either side. The right and left bank dam powerhouses are multi-storied reinforced concrete buildings on the two banks of the Sutlej river,

[2] *International Energy Outlook 2004*, released on 14 April 2004, by the US Department of Energy.

beyond the left and right side retaining wall of the Bhakra spillway. The left bank powerhouse is 122 m (400 ft) long, 33.5m (110 ft) wide and 49 m (161 ft) high, with its lowest elevation at 317 m (1041 ft). Draft tubes and scroll cases were laid over a raft foundation. The concreting was done in two stages to facilitate proper embedment of different equipment. The right bank powerhouse was also of almost similar dimensions, with the lowest elevation at 342.6 m (1124 ft).

The left bank powerhouse had a total installed capacity of 540 MW, comprising five units of 90 MW installed capacity each. The right bank powerhouse had a total installed capacity of 600 MW, comprising five units of 120 MW each. Each generator is of the vertical-shaft, salient-pole type, suitable for the coupling of Francis water turbines, equipped with a thrust bearing and two guide bearings. The power is generated at 11 KV. There are two step-up voltages fixed, that is, 66 KV and 220 KV, for transmission, taking into account the load demand. The 66 KV system would cater to the demands of the Nangal fertilizer factory at a distance of 10 km from the switchyard. The other 220 KV system joins the Nangal grid. One double circuit 220 KV trunk transmission line carries power to Delhi, and another double circuit 132 KV trunk line goes to Ludhiana. It then bifurcates, one going to Jalandhar and the other to Moga/Muktsar.

The left bank powerhouse was constructed between 1959 and 1961. The construction of the right bank powerhouse was initiated in 1964 and completed in 1969. There are many documents that provide complete technical details for those interested in them.[3] Some salient features of the power plants are provided in Table 9.1.

The nation is fortunate that the perseverance of the project sponsors enabled the speedy development of hydropower against some resistance and initial reluctance. The two powerhouses on the NHC at Ganguwal and Kotla, each with two units of 24.20 MW, were the hydropower units, first commissioned in 1955–6. The first unit of the Bhakra dam left bank powerhouse was commissioned in 1960 and the remaining four in 1961. The right bank powerhouse was taken up in 1963 and completed by 1968. Both these powerhouses are interconnected for integrated working and form part of the Northern Regional Grid of India. The final adopted installed capacity of the units not only covers the power potential of the Sutlej waters but also provides for additional power to be generated by the linked scheme—the Beas Sutlej Link Project.

[3] One such document brought out by BBMB itself is *History of Bhakra-Nangal Project*, October 1988.

Table 9.1: Some salient features of the Bhakra power plants

Description	Left bank powerhouse	Right bank powerhouse
Original installed capacity	450 MW	600 MW
Present installed capacity	540 MW	785 MW
Number of units	5	5
Turbines	150,000 HP-Francis type	214,075 HP-Francis type
Rated head	121.92 m (400 ft)	121.92 m (400 ft)
Generator make	AEI England	Kirov Electrosila—USSR
Original capacity	115 MVA at 0.9 PF	134 MVA at 0.9 PF
Uprated capacity	120 MVA at 0.9 PF	174.44 MVA at 0.9 PF
Unit transformers	11/66 KV, 100 MVA English Electric—1 No. 11/66 KV, 120 MVA ABB Make—1 No. 11/220 KV, 135 MVA Russian Make—2 Nos 11/220 KV, 120 MVA Crompton Greaves—1 No.	11/220 KV, 175 MVA National USSR—4 Nos 11/220 KV, 175 MVA TELK Make—1 No.
Overhead crane	Yugoslav 4×115 Tons	Russian 2×250/30/10 Tons

Source: BBMB.

The year-wise generation of power from the BNP since the start of operations has been summarized in Table 9.2.

Performance

Prime Minister Jawaharlal Nehru dedicated the left bank powerhouse to the nation in December 1961, while Vice President V.V. Giri dedicated the right bank powerhouse in April 1969.

The BNP has generated 211 billion kWh of power from 1955–6 till 1999–2000. It continues to generate around 7000 million kWh each year. The Board's achievements, listed by the Council of Power Utilities include highest plant availability (90–4 per cent), highest transmission serviceability (99.5–100 per cent) and lowest cost of generation (6 paisa /kWh).[4]

---

[4] Council of Power Utilities, New Delhi, *Profile of Power Utilities and Non-utilities in India*, 2000, p. 32.

The generating capacity of the powerhouses and units are indicated in Table 9.2.

Table 9.2: Hydro generating units under the BNP

Stations	Unit number	Capacity (MW)	Date originally commissioned	Present capacity (MW)
Bhakra left bank	1	90	14–11–1960	108
	2	90	02–02–1961	108
	3	90	07–07–1961	108
	4	90	06–11–1961	108
	5	90	10–12–1960	108
Total-Bhakra left bank P.H		450	1960–61	540
Bhakra right bank	1	120	24–05–1966	157
	2	120	05–12–1966	157
	3	120	13–03–1967	157
	4	120	13–11–1967	157
	5	120	19–12–1968	157
Total-Bhakra right bank P.H.		600	1966–68	785
Ganguwal on NHC	1	29.25	23–01–1962	29.25
	2	24.20	02–01–1955	24.20
	3	24.20	02–01–1955	24.20
Kotla—on NHC	1	29.25	14–07–1961	29.25
	2	24.20	27–08–1956	24.20
	3	24.20	23–05–1956	24.20
Total—Ganguwal and Kotla		155.30	55–56 & 61–62	155.30
Grand Total Bhakra-Nangal		1205.30	1955–68	1480.30

Sources: BBMB and Council of Power Utilities, New Delhi.

In skewed policy debates, the word 'renewable' is sometimes attached to technologies that are liked and denied to those that are not. This is bad logic though, probably, good politics. Renewable energy is generally defined as one with a fuel source that is not depleted in the production of that energy. Wind, solar, geo-thermal, and hydropower meet that definition. Unfortunately there are powerful competitive lobbies, backed by money power, which would like to see that hydropower is denied the status of a renewable energy.

Table 9.3: Year-wise power generation

(in million kWh)

Year	Power generation	Year	Power generation
1955–6	179	1977–8	5437
1956–7	292	1978–9	6821
1957–8	472	1979–80	6661
1958–9	510	1980–1	5788
1959–60	570	1981–2	6089
1960–1	838	1982–3	6857
1961–2	1586	1983–4	7035
1962–3	2137	1984–5	5932
1963–4	2653	1985–6	5997
1964–5	3005	1986–7	6840
1965–6	3174	1987–8	6345
1966–7	3568	1988–9	7311
1967–8	3656	1989–90	6633
1968–9	4343	1990–1	7630
1969–70	4985	1991–2	7514
1970–1	4281	1992–3	7006
1971–2	4833	1993–4	5752
1972–3	4494	1994–5	7321
1973–4	5812	1995–6	6821
1974–5	3924	1996–7	7276
1975–6	5585	1997–8	5446
1976–7	5714	1998–9	8021
Continued alongside		1999–2000	6954

Source: Bhakra Beas Management Board.

It would be interesting in this context to envisage an alternate scenario, in which this 211 billion kWh of power would have been generated through thermal power plants. The immense adverse impacts on precious natural resources and the environment, are not acknowledged or highlighted. The production of 211 billion kWh using a coal-based thermal power plant would have meant the consumption of 126.6 million tonnes of coal. This would, in turn, have left behind over 40 million tonnes of ash, the disposal of which would have caused immense difficulties. The mining of the enormous extent of coal needed would have also led to the deforestation of a large area. Thus, the energy generated from the BNP has, indeed, saved land and air pollution, thereby preserving and improving the environment, in addition to conserving depleting fossil fuel reserves.

The left bank turbines were among the first large hydro-turbines installed in India. Even during their first two to three years of operation, excessive cavitation was observed on the turbine blades. Therefore, the runner of unit 5 had to be replaced with a new one. However, the damaged runner was taken out and reconditioned by the project engineers at the site. All the runners were also reconditioned in turn, successfully and departmentally.

During the construction period, the river flow had been diverted at the dam site through two tunnels. The hoist chamber of the right diversion tunnel was damaged in August 1959, leading to the flooding of the entire powerhouse building. The stators and governors of units 1 and 2 which were already in position were submerged and remained so for more than two months. It was advised that this equipment might have to be discarded and replacements imported. This would have involved additional costs and time delays. However, a bold decision was taken to use the same equipment after proper cleaning and drying out. This attempt, involving expertise, proved to be a complete success and the units were also commissioned according to the original schedule. They are still serving satisfactorily. Similarly, local expertise was used to handle some other emergencies that arose during these years, such as the damage to the stator core during the phase fault of 1974 and the damage caused by short circuit in 1984. In fact, all capital and normal maintenance and repairs are being carried out in Bhakra-Nangal departmentally, with consequent saving in cost and time.

Renovation, Modernization, and Uprating (RMU)

The project has not only strengthened its expertise in managing the system but has also constantly adopted innovative RMU methods. Such improvements have been a constant feature of the project, ever since it was commissioned.

The right bank powerhouse has five units of 120 MW each, built with the technical assistance of the erstwhile Soviet Union. After carrying out detailed technical studies, these units were uprated in 1980 from 120 MW to 132 MW, by utilizing the in-built safety margins, and without incurring any expenditure. The renovation and modernization undertaken, thereafter, raised the output of each unit to 157 MW. The RMU work commenced in 1995 and was completed by February 2001. The cost thereof was found to be only Rs 5 million against the present day cost of 50–60 million rupees per MW in a new power project. The latest state-of-the-art technology was utilized and the reliability of the plant increased considerably, besides giving it an additional life of two to three decades. The turbine efficiency has also increased by 2 per cent. The peaking capacity thus increased by 5×37 MW equal to 185 MW. With the introduction of a fast-acting static excitation system and modern governing system, the system stability of the grid has improved.

On the left bank powerhouse, there were originally 5 units of 90 MW each. These have been uprated to 108 MW each during the period 1980–5 by the replacement of stator windings. The insulation class of the new stator windings is class F against the original class B type of windings.

Further studies have now been made for the RMU of the left bank power plants, and these have indicated that it would be possible to uprate them from 108 MW to 126 MW each. This work is likely to commence in 2006 and be completed by 2011.

The Ganguwal and Kotla powerhouses each have two machines of Westinghouse make, with an original capacity of 24.2 MW. These were taken up for renovation in the first instance and the work on it has been completed. This involved the modification of the turbine runner, stator, rotor poles and static excitation system, governor and guide vane.

In addition, these powerhouses have one unit each of 29.25 MW—Hitachi machines. In the next phase, the RMU of the Hitachi machines will be taken up. This phase is likely to commence in 2005 and be completed by 2006.

Crucial Contribution to the Northern Regional Grid

For the purposes of planning and operation, the Indian power system has been demarcated into five regional grids, viz., Northern, Southern, Eastern, Western and North-Eastern. Bhakra-Nangal is part of the Northern Regional Grid. The BNP has been playing a crucial role in its smooth operation. Besides providing cheap energy to its participating states, Bhakra is a major source of scarce peaking power. Historically, the Northern Region, like all other regional grids of the country, has been facing a huge peaking power shortage. The present shortage is of the order of 12 per cent.

The power demand rises sharply at the start of the evening peak hours, as lights are switched on almost simultaneously in millions of houses, shops, factories, and commercial establishments. This occurs at any time between 5 pm and 7 pm, depending on the sunset timing, and the peak load lasts for three to four hours thereafter. Similarly, the morning peak load builds up at sunrise from 5 am to 7 am, as activities pick up at homes and in agricultural fields. This peak lasts for a shorter period compared to the evening peak. The Bhakra generators are synchronized one after another during this period, and the load is picked up at a fast rate to match with the sharp rise in load—to help in controlling excrusion of frequency to dangerous levels. Each of the Bhakra generators takes 5–10 minutes from standstill to synchronization, and 1–2 minutes to pick up the full load after synchronization.

Indian grids experience large variations in frequency compared to grids of advanced countries, as only a small number of machines operate in free governor mode for primary regulation, and there are no designated machines for secondary control/Automatic Generation Control (AGC). All the ten machines at Bhakra operate on free governor mode, which has a major contribution in primary regulation of frequency. This helps in stabilizing the frequency of the Northern Regional Grid.

In a grid contingency like the tripping of large-sized thermal machines, over-loading of transmission system, etc., the generation at Bhakra is regulated on the instructions of the Regional Load Dispatch Centre to help in enhancing the grid security and reliability. In the past, a large number of grid disturbances were averted by the timely intervention of the Bhakra powerhouses.

In the event of a grid collapse, 'black start power' is provided by Bhakra. Following a grid collapse, the first Bhakra machine is synchronized within 5–10 minutes, followed by other machines, and start-up power is extended to thermal power stations in Punjab, Haryana, and Delhi. Supply is also extended to essential loads like railway traction, airports, and hospitals.

Bhakra has also been supplying start-up supply to the National Capital Region of Delhi to meet its essential loads following a black-out. On numerous occasions in the past, the supply from Bhakra has been extended to this area in less than an hour. Whenever events of national importance take place in the northern region (like ASIAD-82, the Commonwealth Games, Commonwealth Heads of Governments Meeting), Bhakra plays a major role in the regulation of frequency and in helping the secure and reliable operation of the Northern Regional Grid.

It is difficult to imagine the smooth operation of the Northern Regional Grid without Bhakra and other hydro machines of the BBMB. This has been made possible by the flexibility in operation offered by the Bhakra machines because of a large reservoir.

Transmission System

The BNP transmission system was designed with the dual object of keeping the initial cost as low as possible, and of providing flexibility of conversion at a later date, for the increased transmission of power at a higher voltage when the generation was stepped up. The requirements and possibilities of interconnecting with neighbouring power networks were also kept in mind. The works were carried out by the Punjab PWD-Electricity Branch, the Punjab State Electricity Board (PSEB) or the BBMB at different times. The testing and commissioning of 220 KV lines and sub-stations was carried out departmentally.

The transmission lines/sub-stations constructed under the BNP (including the Bhakra right bank powerhouse) are given in Table 9.4.

Table 9.4: Transmission lines and sub-stations under the BNP

Sl. no.	Transmission lines/sub-stations	Year commissioned
	Transmission lines	
1	66 KV Dhulkote-Chandigarh	1953
2	132 KV double circuit Ganguwal-Kotla	1955
3	220 KV double circuit Ganguwal-Dhulkote-Panipat-Delhi	1955/1962@
4	220 KV double circuit Bhakra (R)-Jamalpur	1968
5	220 KV double circuit Ballabgarh-Badarpur	1968
6	220 KV double circuit Jalandhar-Jamalpur	1969
7	220 KV double circuit Jamalpur-Sangrur	1969
8	220 KV double circuit Sangrur-Hisar	1969
9	220 KV double circuit Hisar-Dadri-Ballabgarh	1970
10	220 KV double circuit Pong-Jalandhar	1978
	Sub-stations	
1	66KV Chandigarh	1953/1962@
2	220 KV Dhulkote	1955
3	220 KV Panipat	1955
4	220 KV Delhi	1955
5	220 KV Jamalpur	1968
6	220 KV Ballabgarh	1968
7	220 KV Jalandhar	1969
8	220 KV Sangrur	1969
9	220 KV Hisar	1969

Source: BBMB.

Note: @ Initially at 132 KV and later upgraded to 220 KV.

An index map of the BBMB transmission network, showing the power stations, sub-stations and transmission lines is given in Figure 9.1.

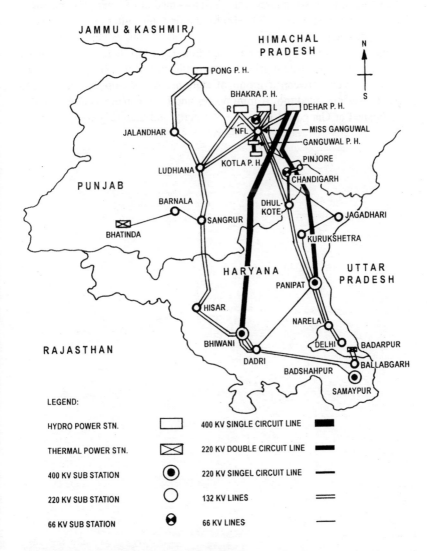

Figure 9.1: BBMB Transmission Network

Communication system

The power transmission lines were used as a medium for carrying messages at high frequencies (50 to 500 kHz) for the first time, after the commissioning of the Ganguwal powerhouse in 1955, under what is popularly known as the power line carrier communication (PLCC) channel. The carrier channels afforded instant speech communication from one station to another, or even between a group of stations. A back-up communication system was provided too. A mini power and load dispatch centre was also established at Ganguwal by using telemetry and metering equipment.

10

FLOOD MANAGEMENT

An Incidental Benefit

A project report invariably indicates, among other things, the various intended purposes and anticipated benefits on the basis of which it is approved for execution. The BNP Report mentions irrigation and power generation as the intended benefits from the project. However, flood moderation was also envisaged as an 'incidental benefit'. The project planning was also done as an irrigation-cum-power generation scheme. Thus, there is no storage space specially earmarked for flood control. The presumption seems to be that there will be savings in the cost by not creating a separate flood space but seeking to derive optimum benefits by the operation of the reservoir through the balancing of conflicting demands.

On the basis of the inflow records of the Sutlej river from 1928 to 1947, and taking the net utilizable storage space at Bhakra with the full reservoir level at 1680 feet as 5.725 million acre feet, the project report predicted that 'the reservoir would not be filled, far less spill, in 65 per cent of the years'.[1] In thirteen of the twenty years (1928-9 to 1947-8), the entire run-off up to Bhakra during the monsoon filling period would have been retained in the reservoir, except for releases to meet the canal demands. In the remaining seven years, the reservoir would fill up and attain a 1680 to 1682 feet level. The project had, therefore, the potential to provide flood control as an incidental benefit.

In the case of any reservoir aimed at flood control, the planning would be to maintain a low reservoir level during high flow months and use the storage capacity to absorb the incoming flood peak(s). Soon after the peak passes down, the reservoir is brought down at a controlled rate to the low level, to be ready for facing the next flood. Reservoir schemes for serving flood moderation purposes alone would not usually be economically

---

[1] *Bhakra Nangal Project*, 1955, vol. I, part I, and Appendix C.

justifiable. On the other hand, multi-purpose reservoirs for irrigation and power benefits could be planned to provide flood moderation as well. If irrigation and power alone were the primary objectives, the endeavour would be to ensure a full reservoir by the end of the filling period. Whenever a storage reservoir is planned for multiple benefits, including flood moderation, the attempt would be to seek the optimum combination of possible benefits through the planned operation of the reservoir, keeping in mind the declared purposes and intended benefits. Under the tropical Indian conditions, a significant part of the annual flow in a river occurs in a period of about three months of the monsoon season but the irrigation, power, and other needs are spread round the year. A rational economic solution to the problem of floods during the monsoon is to link it with the manner of meeting the water demands for irrigation, power and other uses during the non-monsoon season. Thus, an irrigation and hydropower scheme could be made to provide flood moderation during the high flow period and meet various needs by drawing upon the stored water till the onset of the next monsoon.

In retrospect, when Bhakra was planned, great emphasis was laid on drought management in this region and, thus irrigation was emphasized to such an extent that the simultaneous need to tackle floods was sidelined. There is now a widespread recognition that the time has come for making a paradigm shift in the manner we view drought management and flood management. In arid and semi-arid tropical regions, drought management and flood management are really two sides of the same coin. Drought management would need to be intertwined with flood management, so as to achieve multiple benefits with a single intervention.

When a conservation reservoir is called upon to serve a flood moderation purpose, planned or unplanned, the operating authority is often faced with a difficult choice, particularly if the incoming flood peak occurs towards the end of the filling season. If the high reservoir level were brought down, based on inflow forecasts and best judgment, the authority could still be criticized if the reservoir did not fill up later, leading to shortfalls in meeting needs. The authority risks equal criticism if the reservoir were not sufficiently lowered to absorb the incoming peak and this resulted in flood damages lower down. Constituencies that have been lobbying against 'large dams', state that 'man-made floods' could be caused by them. Such a criticism might even sometimes acquire an edge with the aid of hindsight which, unfortunately, the operating authority does not enjoy. There have been numerous instances of such dilemmas encountered by project authorities in India.

Flood Management by Bhakra

The BNP authorities claim that, irrespective of whether it was an intended purpose or not, the fairly large storage capacity of the Bhakra reservoir was always utilized in such a manner as to render flood moderation benefit for the region below the dam. This is quite understandable and has, indeed, been the case with many a large dam in India and elsewhere. In fact, there are hardly half a dozen dams in India where some specific flood storage spaces are reserved for flood control. Many of the multi-purpose reservoirs without allocated flood reserve, particularly the larger ones, have rendered significant flood moderation benefits.

This chapter examines, with the data available on reservoir operations since the project was commissioned and the benefit of hindsight, the validity of the claim made and the extent of flood moderation actually achieved, as also failures, in this regard, if any.

River Channel

After entering India, the Sutlej river takes a south-westerly direction on its way to the Bhakra gorge, about 322 km away. The river leaves the Himalayas near Nangal and enters the Anandpur Dun, a valley and plain area between the outer range of the Himalayas and the Siwaliks. This valley extends north-south from Nangal to Kakrala near Roopnagar for about 50 km in length, with an average width of 10 km. Soon after Nangal, the Soan *nadi* (river) joins it from the right side and, thereafter, the Sirsa nadi from the left. The river has a braided course here, with seasonal torrents and tributaries joining it. It emerges into the plains near Roopnagar. Several streams join it between Roopnagar and Ferozpur.

Before the construction of the Bhakra dam, the channel downstream of Bhakra was in the form of flood plains, including the flood channel and the adjacent marshy lowland. The flood plain was about 7 to 8 km in width. The river constantly shifted its course during floods. A large extent of land was also inundated during high floods. Besides, the river had a tendency to meander freely from one bank to the other.

Canalization and Embankment of Downstream Reaches

According to past records, the highest flood discharge which had passed down Roopnagar was around 490,000 cusec (13,875 cumec). After the dam was constructed, it was expected that in most years the major component of the flood discharge up to Bhakra would be stored in the reservoir. To make beneficial use of the flood plain in the post-Bhakra dam situation, the Government of Punjab decided to canalize the river. A project was prepared for the canalization of the Sutlej by embankments on both banks spaced 3000 to 3500 ft (914 to 1067 m) apart. The canalized section was stated to have been designed to cater to the moderated flood discharge of 200,000 cusec (5663 cumec).[2] The embankments would extend from Ropar to Giderpindi, near the Harike barrage. Thereafter, continuous flood embankment was provided only on the left bank till Hussainiwala. This flood embankment was designed on the basis of the last observed high flood level.

The design discharge of 200,000 cusec (5663 cumec) has, however, been a subject of controversy ever since. Two committees, one set up by the state government and another by the central government, had re-examined the design discharge and other features of the canalized section. As the present study is on the impact of the BNP alone, the canalization and embankments will be touched upon in a limited sense.

After the commissioning of the Bhakra dam in 1963, floods in the first few years of its operation were absorbed in the reservoir. For the first time, water spilled over the Bhakra spillway in 1973. This was also the first time that the embankments were really subjected to a performance test. Thereafter, the floods of 1975, 1978, and 1988, put them to further severe tests. The 1973 flood saw many breaches in the canalized reach on both banks. In 1975, high flood water levels in the canalized reach almost rose to the top of the banks but overtopping was averted. The 1978 and 1988 floods were full-scale tests when problems were highlighted. The results thereof are examined subsequently in this chapter.

It will be apparent that when the flood impinges into the reservoir in July or August, the water level in the reservoir is likely to be lower than the FRL. However, if the flood occurs in October or late September, a difficult judgement will need to be made by the operating authority. The extent of reliable inflow/flood forecasting arrangement that exists, and the type of reservoir operation instructions available, would influence the decision. Whatever the precision of these aids, the situation would call for sound judgement, based on past experience, by the regulatory authority.

A study by the WAPCOS using the past data of high flows, indicates that in the Sutlej basin up to Bhakra, peak floods are most likely to occur in the months of August and July, in that order.[3]

---

[2] It has been stated by different people that as the embankment stands today it will not be able to take 200,000 cusec, or even 100,000 cusec.

[3] Bhakra Beas Management Board, A study made by WAPCOS, 'for official use', 1996.

The month-wise probabilities are as indicated in Table 10.1.

Table 10.1: Month-wise Count of Annual Maximum Floods
(1909–95)

Month	Number of annual maximum floods
June	2
July	35
August	44
September	5
October	1
Total	87

Effect of the Reservoir on High Flows

One possible way of evaluating the effect of the Bhakra reservoir on flow moderation would be to consider the daily average inflows and outflows during the period since the dam was commissioned. An analysis of the data for the high flow months of July, August, and September during the years 1967 to 1995 was done. It was found that there were seventeen occasions when the daily average inflow was greater than 100,000 cusec. The corresponding outflow on fifteen of these occasions was less than 100,000 cusec. However, on two occasions, namely, on 20 August 1978 and 25 September 1988, the average daily outflow exceeded 100,000 cusec. The details are presented in Table 10.2.

A comparison on the daily average basis of inflows and corresponding outflows, as has been shown in Table 10.2, might mask the effective moderation of the peak flood. Perhaps hourly or three hourly average flows, if compared, could indicate the extent to which the flood peak was cut to a moderate flow. However, comparing the hourly inflow and outflow, too, has its demerits. Soon after any peak inflow recedes, the stored flood water would need to be released in a controlled manner, in accordance with the reservoir regulation manual. This would enable the reservoir level to be lowered, so as to face the next possible flood peak. Such lowering can take place only if the outflow is more than the inflow.

The intensity of a flood is generally indicated by the magnitude of instantaneous peak discharge. This figure is needed in any case for design, as well as for the constant evaluation of the safety of structures and the downstream embankments. Another relevant parameter is the volume of the

Table 10.2: Moderation of daily average flow by the Bhakra Reservoir during July–September, 1967–95

(All flow figures in 100,000 cusec)

Month and date	Year	Inflow (daily average)	Outflow (daily average)
26 August	1967	1.2	0.10
11 August	1969	1.0	0.15
7 August	1971	2.0	0.10
27 July	1973	1.1	0.20
27 July	1976	1.1	0.20
5 August	1977	1.1	0.20
3 August	1978	1.0	0.20
20 August	1978	1.1	1.10
14 July	1980	1.0	0.20
21 August	1985	1.1	0.20
8 July	1986	1.2	0.22
9 August	1988	1.1	0.36
25 September	1988	1.1	1.20
1 August	1989	1.2	0.20
14 August	1990	1.1	0.30
21 July	1994	1.2	0.25
6 September	1995	1.1	0.53

Source: *BBMB/WAPCOS*.

flood, as flood moderation occurs by storing a part of the flood inflow in the reservoir.

The design flood hydrograph of the Bhakra dam indicates the peak inflow as 400,000 cusec, the flood duration is of 90 hours and its volume is 1,788 MAC ft. The space available above the FRL would not, therefore, suffice to effectively manage the design of the flood volume, if such a flood arrives when the reservoir is nearly full. Dam safety considerations will not then permit any significant flood moderation. This would be particularly so where irrigation and power are the intended benefits of the project and flood moderation is an incidental benefit.

It was indicated that the designed channel capacity of the river-reach downstream of the dam is 200,000 cusec, whereas that of the Bhakra spillway is 290,000 cusec. Many tributaries also join the Sutlej downstream. These by themselves are said to be capable of adding 200,000 cusec of flow during heavy rains. The peak flood discharge on 26 September 1947, was 275,754 cusec at Bhakra, and 490,000 cusec at Ropar. Thus, the discharge that was added in the reach from Bhakra to Ropar was 214,246 cusec. Therefore, the design discharge adopted for the embanked reach lower down, involved a conscious decision with a calculated risk. Usually, embankments protecting agricultural rural areas are designed to be safe enough for a flood with a return period of once in twenty-five years. In other words, such embankments will not be able to withstand very high floods of probability of occurrence higher than one in twenty-five. The Sutlej embankments are of this type.

The project considers the period from 21 May to 20 September as the reservoir filling season and 21 September to 20 May as the depleting season. During the depleting period, the releases are generally in accordance with the irrigation and power demand (varying from 20,000 to 28,000 cusec).

Upstream Flash Floods

The reaches of the Sutlej upstream of Bhakra are prone to flash floods. Sixty-five per cent of the catchment of the river up to Bhakra lies in the Tibet region of China. Often, information about flash floods originating upstream does not reach downstream areas till the flood itself arrives. Hence, they often cause loss of lives and property in India. Upstream flash flood is caused by local cloudbursts and/or by temporary blockage of the river, due to landslides and the breaching thereof, leading to onrush of water. The worst such flash flood occurred in the year 2000.

Between midnight of 31 July and early morning of 1 August 2000, an unprecedented flash flood arrived in the Kinnaur district of Himachal Pradesh. This flood originated in Tibet and entered India around 0100 hours. It reached the Nathpa-Jhakri Project (NJP) near Rampur, at around 0400 hours. The Bhakra Dam Organization has a discharge observation site at Rampur.The records of the Rampur site indicate that the discharge shot up from 1535 cumec at 0300 hours to 5097 cumec at 0530 hours. By 0900 hours the flood peak tapered off. According to the NJP authorities, the water level of the Sutlej rose by 15 metres within a short span of time. It entered the NJP powerhouse through the tail-race channel and caused immense damage to the power equipments. Loss of lives and property of the nearby areas were also reported.

This flash flood was fully absorbed and evened out by the Bhakra reservoir on 1 August 2000. The release downstream of the Nangal barrage remained unchanged at a very low level of about 650 cumec (23,000 cusec) throughout that day. During this period, the reservoir level in Bhakra rose by 1.7 metres. In the absence of the Bhakra dam and the large reservoir behind it, the flood could have travelled downstream to the Punjab plains and caused devastation. Thus, the Bhakra dam has been very effective in mitigating the effect of flash flood originating upstream, too.

Actual Operation and its Impacts

Based on the actual operations carried out till now, a study has been made of the flood moderation by the Bhakra dam during the period 1964–2002, for all inflows greater than 200,000 cusec. The actual operations and their effect, based on hourly flows, are presented in Table 10.3.

Flood Regulation After the Peak Passes

After moderating the peak flood inflow, the reservoir level may need to be brought down so that it would be possible to regulate likely high inflows during the subsequent period of that flood wave, or any subsequent flood. The past releases downstream of Nangal were examined to see the number of occasions (after the peak flood inflow has already passed) in which the releases downstream were higher than 100,000 cusec. It was again found that in 1978 and in 1988 there were such occasions when releases higher than 100,000 cusec had been made.

In 1978, the peak inflow of 378,769 cusec reached Bhakra on 19 August at 0320 hours, when the reservoir level was 1682.19 feet. The corresponding total outflow downstream of Nangal was only 54,950 cusec. However, the maximum outflow made later that night, between 2400 hours on 19 August and 0100 hours on 20 August, was 137,356 cusec. This was slightly higher than the corresponding inflow at that time but was done to keep the reservoir at a safe level to meet likely incoming floods in the remaining monsoon months.

In 1988, the peak inflow of 318,182 cusec reached Bhakra on 26 September at 1600 hours, when the reservoir level was approaching 1686.45 feet. The corresponding total outflow from Nangal was 131,260 cusec. However, the maximum outflow made at 1700 hours that day, was 148,740 cusec. Releases of over 100,000 cusec continued to be made for twelve more hours, till 0500 hrs on 27 September.

In contrast, a larger flood that occurred in 1971 was moderated very significantly because the reservoir level was very low at that time. The flood inflows recorded were higher than 200,000 cusec between 1000 hours and 1700 hours on 6 August 1971. The peak inflow of over 608,383 cusec,

Table 10.3: Moderation by the Bhakra Dam

(all discharge figures are expressed in cusec)

Year	Date	Hour	Inflow	Outflow
1969	23 July	0800	222,771	15,600
1971	6 August	1000	256,549	11,250
		1100	292,562	11,000
		1200	397,840	10,750
		1230	608,383	10,500
		1300	502,719	10,500
		1400	384,377	10,350
		1500	363,971	10,200
		1600	265,611	10,050
		1700	237,359	9,900
1973	26 July	0600	205,495	20,450
		0700	208,272	18,487
		0800	170,886	18,487
		0900	203,686	18,487
		1000	233,349	18,487
		1030	271,826	18,487
		1100	228,612	18,487
1975	4 September	0700	230,096	58,220
1978	19 August	0300	176,218	58,860
		0320	378,769	54,950
		0400	263,550	54,950
1981	14 July	1000	143,036	18,000
		1030	268,148	18,000
		1100	205,351	15,000
1985	20 August	0800	212,504	12,000
		0815	245,784	12,000
		0900	143,237	12,000
1988	26 September	1500	211,086	124,904
		1600	318,182	131,260
		1700	254,832	148,740
1990	13 August	0700	233,464	25,000
		0800	167,175	25,000
		0900	253,066	30,000
1992	28 August	0400	234,101	30,000
1994	20 July	0700	224,859	27,000
		0800	206,325	27,000
		0900	209,919	27,000
1995	14 August	0700	220,990	12,500
		0800	285,414	12,500
1997	12 August	1100	202,252	17,500

Source: Records of the BBMB-Bhakra Dam Organization.

which was much higher than the peak of either 1978 or 1988, was recorded at 1230 hours on 6 August 1971.The outflows were all less than 11,250 cusec.

Flood Forecasting and Warning System

The BBMB maintains a network of gauge and discharge stations and meteorological observatories established along the periphery of the reservoir and the catchment area. These include rainfall and snow observation sites. Three river gauge stations on the main river upstream are at Rampur, Suni, and Kasol; and six stations are on *khuds* (tributaries). Discharge observations are made daily in the monsoon and twice a week in the non-monsoon season. The Olinda site on the Sutlej is downstream of the Bhakra dam.

During the monsoon season, the project's regulation system keeps in close touch with the IMD. Information on cloud cover in the catchment, quantum of rainfall, weather conditions, etc. are regularly collected and analysed.

Wireless sets have been set up at Rampur, Kasol, Suni, Kahu, Barthin, and Nangal. These are further linked with Chandigarh and Delhi. Control rooms are at Bhakra and Nangal. A regulation cell functions under a director at Nangal. The regulation division regulates releases from the reservoir to meet the flood situation, besides releases for irrigation/power. It also disseminates flood forecasts. When the releases from the Nangal barrage are likely to exceed 50,000 cusec (1416 cum.), flood warnings are issued to the Central Flood Control Room, the chief engineer, Punjab, and the deputy commissioner, Roopnagar district.

Experts and consultants have made suggestions from time to time for the improvement of the existing flood forecasting system. The importance of flood forecasting and warnings, as also the need for periodic updating of the system, should never be underestimated.

Flood Problems in Downstream Reaches

It was noted that under the existing procedures, releases up to 50,000 cusec could be made below the Nangal barrage without any warning to local civilian authorities, or to the general public. Another 8000 cusec could be passed into the Bhakra mainline canal. It would seem that releases of the order of 58,000 cusec below Bhakra are acceptable without any problem. Table 10.3 showed that this limit had been exceeded in the year 1988.

Records indicate that, on many other occasions, after the peak flood had passed down, the outflows were higher than the releases made at the time of the peak inflow. This is understandable, as the reservoir level needed to be brought down soon after the flood management operation. This was the case in 1992, 1994, and 1995 but releases in these years were less than 60,000 cusec. This would not cause problems lower down. In fact, if the releases were necessary in the interest of safety, this could be extended to some safe limit, say 100,000 cusec. Judged in this light, in 1978 too, the release downstream of Bhakra crossed 100,000 cusec and touched 137,356 cusec, though, as was noted earlier, this was *not* at the time of the peak inflow.

Barring these two instances of 1988 and 1978, the Bhakra dam was able to moderate all flood peaks, restricting the outflow to 100,000 cusec, if not much lower than that. Therefore, the specific circumstances in 1978 and 1988 were examined.

1978 FLOOD

At Ropar, a flow of the order 110,000 cusec was experienced on 2–3 August 1978. The additional flow from torrents on both banks downstream of Ropar swelled the river. At Phillaur, the high flood level was exceeded by 3.5 ft (over a metre) on 3 August 1978, causing overtopping and resultant breaches. The Bhakra spillway was opened only on 7 August 1978, and the releases from the Bhakra dam were still around 37,550 cusec. There were heavy rains in the catchment downstream on 18 and 19 August. Meanwhile, the outflow from Bhakra, too, increased and crossed 100,000 cusec on 19 August 1978. The combination of these circumstances caused floods downstream and extensive damage to the embankments along the Sutlej, from Roopnagar to Harike. In its wake, two technical committees were set up, one by the Government of Punjab and the other by the Government of India. The Government of India Committee analysed the 1978 peak flow and stated that:[4]

The discharge experienced in 1978 was estimated to be 256,282 cusec at Ropar at about 18 hours on 19–8–78. The peak discharge from Gobind Sagar reservoir measured at Olinda was 107,248 cusec at 22 hours on 19–8–78. However, the time lag between Bhakra dam and Ropar has to be taken into account. This is said to be between 12 and 24 hours. According to the hourly statement of discharges

[4] Report of the Committee of Technical Experts regarding the causes of breaches in the Sutlej river and permanent solution of the problem, November 1978.

through the outlets and Spillway furnished by the Bhakra and Beas Management Board to the Department of Irrigation, it is seen that the outflow from Bhakra dam between 18 hours on 18–8–78 and 6 hours on 19–8–78 was of the order of 60,000 cusec. The peak of the flood originating in the catchment between Bhakra dam and Ropar (about 1071 sq. miles) can, therefore, be taken as 256,282–60,000 = say 200,000 cusec.

The Committee had highlighted the fact that the extent of flow generated in the catchment downstream of the Bhakra dam was significantly more than the regulated release from the dam. It also made a number of recommendations on raising and strengthening the embankments, the revised design of embankments, review of the Bhakra reservoir regulation by the BBMB, flood, inflow forecasting, maintenance and so on.

Most of these recommendations, however, remained unimplemented for various reasons, till the next high flood occurred in 1988.

1988 FLOOD

The Sutlej flood of late September 1988 occurred after the end of the filling season and with a full reservoir storage. The monsoon of 1988 had been good and the reservoir stood at a level of 1687 feet on 20 September. The inflows suddenly started rising from 23 September and crossed 100,000 cusec by 20 hours that day. The peak inflow rose to 318,182 cusec at 16 hours on 26 September. Thereafter, it started subsiding.

A well-marked low pressure area, which produced very high rainfall, resulted in unprecedented floods. Rainfall analyses showed that intense rainfall during that rainstorm was widespread and well distributed. However, it was concentrated near Nawanshahr in the middle of the catchment. At Bhakra dam (Khosri camp), rainfall for the 24 hours ending 0830 hours on 24 September 1988 was 281.5 mm. The volume of flood flow that occurred between 23 and 27 September was about 0.85 MAF. With the reservoir standing at over 1687 feet on 23 September, it was not possible to store any more inflow. The best that was possible was to maintain the outflow to match the volume of the inflow, however, restricting the peak outflow at 148,740 cusec in order to minimize the losses downstream. It so happened that the area downstream, too, received high rainfall at that time. The flow at Ropar was estimated at 478,000 cusec. This had happened because of the flooding of 58 torrents that meet the Sutlej river downstream of Nangal barrage but upstream of the Ropar headworks. Two principal torrents of this category are the Swan and the Sirsa. The corresponding inflows in the Swan and Sirsa rivers alone accounted for up to 238,000 cusec.

At Phillaur, it was assessed at 630,000 cusec. There were many breaches in the embankments and people suffered. The Punjab Government reported that over four million people suffered, nearly 2.8 million ha crop area was affected, and that the state suffered colossal damage to crops, houses, and public utilities.

The late September 1988 flood was the fifth in a series that year, preceded by four flood events when the inflow exceeded 100,000 cusec. For the same reason, the reservoir was full at the fag end of the filling season when this high flood occurred. The reservoir could not, without jeopardizing the safety of the dam, accommodate the additional flood inflow. Nevertheless, the reservoir operation effected a moderation of the flood. It was unfortunate that at the same time, the lower region below the dam was equally in spate due to high rains. In the absence of the Bhakra dam and reservoir, the peak flows would have passed down the Bhakra gorge completely unmoderated. With the Punjab state itself in turmoil, the added agony of flood losses led some critics, with the subsequent advantage of hindsight, to find fault with the operations.

The most serious outcome was the charge made by some critics that the BBMB caused the disastrous floods by 'opening the flood gates of Bhakra and Pong.'[5] On that pretext, some Khalistani militants killed the then BBMB chairman, Gen. B.N. Kumar. Available evidence shows that the flood damage would have been much more but for the moderation provided by the dam. Under the existing circumstances, the serious accusations made against the regulating authorities were unwarranted. Some critics have, however, continued to voice complaints. Many others have examined the situation carefully, based on data and analysis, and have concluded that the Bhakra reservoir 'admirably performed the role of flood control in spite of this being only an incidental benefit'.[6]

Post-1988 Review

After the tragic events of 1988, the central and state governments, which participate in the BBMB have made an elaborate review of the functioning

[5] Verghese, B.G., *Winning the Future*, Konark Publishers Pvt. Ltd, Delhi, 1994, p. 23.

[6] This quote is from the WAPCOS study, which added that there is absolutely no question of the construction of the reservoir having resulted in an increase in the flood risk. B.G. Verghese (ibid.) said that 'the project has been an astounding success' and that: 'were it not for the cushion provided by the two dams, the floods would have been even worse'. Many others have voiced similar views.

of the project, keeping the flood management aspect as the main focus. The services of expert consultants were also availed of in this task.

The BBMB in July/August 1990 decided on some fresh guidelines for the Bhakra reservoir as follows:

The rule curve of the Bhakra reservoir would be modified with effect from 1 September 1990. The reservoir would not be allowed to be filled:

– Beyond 1650 feet by 31 July
– Beyond 1670 feet by 15 August
– Beyond 1680 feet by 31 August

It was also decided that the maximum level at Bhakra should be kept at elevation 1680 feet for storage purposes. However, in case the levels were allowed to rise a little higher for the purpose of flood routing/absorption, and to avoid synchronization of releases with those of rivulets downstream of the dam, the level would be brought down as soon as conditions downstream of the reservoir permitted. Similarly, corresponding decisions regarding the Pong dam were also taken.

The consultant for the BBMB, commissioned for a study on these issues, had recommended in 1996 that flood control should no longer remain an incidental benefit for Bhakra. It should be made one of the main benefits for which provision must be made. The full reservoir level should be, therefore, 1680 ft. and the storage space between 1680 ft and 1686.5 ft should be used for flood moderation. An improved inflow forecasting system and improvements to the carrying capacity of the river channel downstream of Ropar, were also recommended.

Expert Recommendations

Perhaps the BBMB is still awaiting the emergence of a consensus among the participating states in these matters. Evidently, the stakes in flood moderation are the highest for Punjab; the downstream states, Haryana and Rajasthan, would prefer higher conservation storage. The same clash of interest between flood control and irrigation/power are reflected in the differing interests of these three states. A statesman-like resolution of these issues is in the best interests of all. There certainly cannot be any lasting advantage in creating economic prosperity through greater irrigation and power infrastructure, if floods are allowed to wipe out the accrued gains once in every few years. The signs for the future seem promising, as even the partial 1996 decision to improve the flood moderation aspect is a rightful recognition of the hard realities. However, a note of caution must be struck that a quick decision needs to be taken on the lines of the recommendations that emanated from the post-1988 review. A change in the reservoir rule curve, keeping the maximum level no higher than 1680 feet for storage purposes, is one on which the decision cannot be postponed any longer. In its absence, should a high flood of the type of 1978 or 1988 occur again towards the end of the monsoon season, significant losses downstream cannot be avoided.

There are some good case studies of the performance of multi-purpose reservoir schemes, with and without specific flood reserve allocation, with respect to flood management.[7] These have generally underlined their effective performance but have stressed that there could be a clash of interest between flood control and other uses like irrigation and power generation.

The Planning Commission's 'Working Group on Flood Control', set up for the Tenth Plan (2002–7), had recommended the review of 'the major dams in India to suggest if the reservoir regulation can be modified for better flood management in the future'.[8]

The Experts Committee on the Review of Implementation of Recommendations of the Rashtriya Barh Ayog (National Flood Commission), more popularly known as the Rangachari Committee (2003), 'noticed a tendency to dilute the importance given to the flood control aspect, by giving greater importance to hydropower and irrigation'.[9] It held that the reservoir regulation schedule could be vastly improved through a revision in existing procedures and guidelines by giving flood management its due importance. It is hoped that the government will soon consider making flood management one of the principal objectives of the BNP and modify the reservoir regulation instructions accordingly.

Without Bhakra?

It was noted that the Bhakra reservoir was able to moderate all peak flows that passed through it in the last four decades, in addition to rendering

[7] For instance, see the study on DVC reservoirs by the National Flood Commission (1980) and the cases included in the article presented at the Ninth National Water Convention, November 2001, by R. Rangachari on the *Role of Storage Dams in the Management of Floods*.

[8] Report of the Working Group on Flood Control Programme for the Tenth Five Year Plan—chaired by R. Rangachari , Chapter 5, para 5.3.5 and Chapter 16, para 16.6.3, Government of India, August 2001.

[9] Government of India, Ministry of Water Resources, *Report of the Experts Committee on the Implementation of the Recommendations of the National Flood Commission,* April, 2003.

significant irrigation and power generation benefits. It was further concluded that without the Bhakra dam in position, a disastrous flood like the ones that occurred in 1978 or 1988, would have passed down completely unmoderated, and that greater harm would have been done to lives and economic prosperity. Another possible perspective is that if the BNP would not have been undertaken, perhaps the canalization and embankments might also not have been undertaken. In such a case, there would have been negligible economic activity of the type that now exists in the flood plains of the Sutlej river. Consequently, there might have been nothing of real worth for the Sutlej river in spate to destroy in its flood plains. However one views the picture, there can be no denying the fact that the prosperity of the region is due to the BNP. The best manner of sustaining and enhancing this prosperity would be to improve the overall performance of the project in all possible ways.

11

INCIDENTAL AND INDIRECT BENEFITS

Fruits of the Project

The BCB examined the project report of 1951 and directed that it should be revised in certain respects. This was done in 1953. The project report and estimates were then considered and approved by the BCB in 1955. The report stated that this multi-purpose project would provide water for irrigation and cheap hydroelectric power. Incidental flood control benefit was also anticipated. It was held that there was no serious flood control problem in the Punjab but the diminution of floods would enable reclamation of some riverine land in narrow strips along the Sutlej river. The principal benefits remained unchanged in the revised 1955 Report. The Report sums up the 'Fruits of the Project' by not only mentioning its principal objectives but also listing various incidental benefits. These were:
1. Prosperity of the agricultural community and immunity from famines
2. Increased production of food crops and cash crops
3. Industrial development
4. Reclamation of state wastelands
5. Rehabilitation of refugees

Certain other benefits were indicated at different places and times. Some other benefits, too, have actually accrued. In the earlier chapters, all the principal benefits have been discussed. Flood management, irrigated agriculture and increased prosperity, and increased food production, which are incidental objectives, have also been discussed. The various other indirect and incidental benefits are discussed below.

Incidental Benefits

IMMUNITY FROM FAMINES

The area brought under irrigation by the BNP is 2.7 million ha (6.6 million acres) in Punjab, Haryana, and Rajasthan. In addition, irrigation water supplies for the already existing Sirhind canal and Western Jamuna canal areas were improved. The area served by the BNP had often been subject to the failure of rainfall and resultant famines. The First Irrigation Commission (1901–3) stated that in a period of 50 years, south-east Punjab was subject to thirteen dry years, which included five years of severe drought.[1] The government was forced to spend vast sums of money on famine relief measures. Whenever the rains failed, the scarcity of food grains and drinking water, and the migration of people and animals were common. After the drought of 1896–7, the Second Famine Commission was set up. It recommended that 'among the measures that may be adopted for giving India direct protection from drought, the first place must unquestionably be assigned to irrigation'. In fact, the appointment of the First Irrigation Commission was itself a sequel to two famines which had highlighted the need to extend irrigation as a protection against future famines.

The First Irrigation Commission made a strong plea to keep the waters of the Beas and Sutlej free for irrigating the areas on the left bank of the latter (i.e., southernmost Punjab, Rajputana), instead of utilizing them for the Lower Bari Doab canal on the right bank. This is what the BNP proposed to do. By the time the BNP was completed and irrigation commenced, it coincided with the ushering in of the new strategy in India for irrigated agriculture. The new strategy, popularly called the 'Green Revolution', anchored by controlled irrigation with the conjunctive use of the BNP canal and groundwater, and HYV seeds, entirely changed the picture of agricultural production in Punjab and Haryana. Those unfamiliar with the past picture of Punjab would find it incredible that this region, which in the last five decades has served as the granary of India, was ever the scene of droughts and devastating famines.

The most recent experience of widespread drought in India was in 2002, when the monsoon (July–September) rains failed. July, which normally gets 30 per cent of monsoon rains, was the driest with a 50 per cent shortfall, a record low in the last hundred years. The total monsoon rainfall was only 81 per cent of the long-period average. Similar past widespread failures occurred in 1987 and 1980. Even under such difficult circumstances, the resilience imparted to agricultural production owing to the extensive irrigation, in this region and the improved economy was clearly evident.

POVERTY ALLEVIATION

A high level of poverty is synonymous with poor quality of life, deprivation, and low human development. The Planning Commission periodically

---

[1] Report of Irrigation Commission, 1903, para 15.

estimates the incidence of poverty at the national and state level, and such records for the period 1973–4 to 1999–2000 are available. Based on the large-scale sample survey on consumer expenditure, the poverty ratio was estimated at 27 per cent in rural areas and 23.6 per cent in urban areas, aggregating to 26.1 per cent for the country as a whole. At the national level, there has been a decline from 1973–4 up to the present but rural-urban and interstate disparities are visible. Regions that experienced a faster rate of output growth (GNP), particularly where such growth was mainly through agricultural production, exhibit reduction in the proportion of people below the poverty line. Punjab and Haryana indicate the lowest levels of poverty, while Orissa and Bihar are at the other end. Furthermore, there are no marked urban-rural variations in Punjab and Haryana. Their positions are shown in Table 11.1.

The availability of assured irrigation and plentiful electricity for house-holds, industries, and agriculture, in which the Bhakra dam played a very crucial role, had thus helped in bringing down the poverty level in the states of Punjab and Haryana. The benefits of the economic development in this region have been shared both by rural and urban areas.

A well-targetted and properly functioning PDS is important for poverty alleviation. The PDS has evolved as a major instrument of the government's poverty eradication programme and is intended as a safety net for the poor. We have already noted that Punjab and Haryana play a key role in the central pool of procurement of food grains and in India's public distribution system. A part of the centrally procured food grains is provided to the states during emergencies to run the 'Food for Work' programme, where a part of the wage payment is made through food grains.

RECLAMATION OF STATE WASTELANDS

It was noted earlier that in the pre-dam period, the river channel downstream of Bhakra spread out and became a fairly wide flood plain. The Sutlej flood plains were generally 7 to 8 km wide and were largely uninhabited due to the high flood risks. After the Bhakra dam was constructed, the Punjab Government planned to make use of the flood plain and built embankments along both banks. The narrowly canalized embanked reach of the Sutlej was designed to carry a discharge of only 5663 cumec (200,000 cusec). The land outside the embankments was soon occupied and put to intense use, both for agricultural purposes and urbanization. This canalization and reclama-tion of land had proved quite useful against flash floods for many years but came under distress in years of sustained high floods, as discussed earlier in the chapter on flood management.

Table 11.1: Poverty ratio

(%)

State	Rural			Urban			Combined		
	1973–4	1993–4	1999–2000	1973–4	1993–4	1999–2000	1973–4	1993–4	1999–2000
India	56.44	37.27	27.09	49.01	32.36	23.62	54.88	35.97	26.10
Haryana	34.23	28.02	8.27	40.18	16.38	9.99	35.36	25.05	8.74
Punjab	28.21	11.95	6.35	27.96	11.35	5.75	28.15	11.77	6.16

Source: Economic Survey 2001–2, Government of India, Ministry of Finance.

The failure of the embankments in many reaches, particularly during the high floods of 1978 and 1988, had led expert committees to examine the position and culminated in many recommendations to protect the reclaimed areas from future flood damage losses. These recommendations have, however, hitherto remained largely unimplemented. It is time these were examined urgently and the necessary action taken.

REHABILITATION OF 'PARTITION' REFUGEES

Millions of refugees suddenly arrived in India in the wake of the Partition of India at the dawn of Independence. Estimates of the scale of migration in either direction vary. The geographer, O.H.K. Spate, puts the total at 17 million.[2] V.P. Menon, puts it at 5.5 million each way in the West.[3] R. Jeffrey arrives at a figure of 12 million for Punjab.[4] Perhaps we could consider that 6 million or so came from West Pakistan. Dr M.S. Randhawa noted that 'it was felt that only an ambitious multi-purpose project like the Bhakra dam could help to mitigate the suffering among the refugees and to solve a rehabilitation problem of gigantic dimensions'.[5] The rehabilitation of the large number of cultivators who possessed vast experience of irrigated agriculture, as also the many artisans and small-scale industrialists in the region served by the BNP, has been chronicled by many. The quick settling down of the refugees who migrated to India from West Pakistan (in comparison with the somewhat slower pace in respect of those from East Pakistan), owes a lot to the irrigation and power benefits offered by the BNP. It has, indeed, proved to be the single most effective step taken by the government for their rehabilitation.

LIVESTOCK AND MILK PRODUCTION

One way of raising productivity from land involves ruminants like cattle. They have a digestive system that enables them to convert roughage which humans are unable to digest into animal protein. This ability of theirs can convert crop residues like straw and stalks into meat and milk. India has been steadily increasing the production of milk over the last four decades.

---

[2] Spate, O.H.K., *India and Pakistan: A General and Regional Geography*, Methuen, London, 1954.

[3] Menon, V.P., *The Transfer of Power in India*, Orient Longman, Bombay, 1968.

[4] Jeffery, R., *Modern Asian Studies*, VII 4, 1974.

[5] Randhawa, M.S., *A History of Agriculture in India*, Indian Council of Agricultural Research (ICAR), New Delhi, 1986, vol. 4, Chapter 6, p. 80.

She overtook the USA as the world's leading milk producer in 1997. Between 1961 and 2001, India's annual milk production went up from about 20 million tonnes to 85 million tonnes. Punjab, Haryana, and Rajasthan have played a leading role in India's milk production. Since the early 1970s, about one-sixth of India's production has come from Punjab and Haryana alone. Besides, they stand first and second in per capita milk production, right from 1971. Milk production in the main states of the BNP region is indicated in Table 11.2.

Table 11.2: Milk Production in India

(Production in million tonnes
Per capita production in kg/annum (rounded off))

Year	India		Punjab		Haryana	
	Pro-duction	Per capita production	Pro-duction	Per capita production	Pro-duction	Per capita production
1966–7			1.9	137	1.1	128
1971–2	22.5	41	2.1	157	1.5	151
1981–2	34.3	50	3.5	207	2.3	176
1990–1	53.9	63	5.1	249	3.4	207
1992–3	57.9	66	5.6	269	3.7	218
1997–8	72.1	74	7.2	309	4.4	227
1999–2000	78.8	78	7.7	320	4.7	230

Source: *Statistical Abstracts of India, Haryana and Punjab for different years CMIE, Mumbai, Basic Statistics, States, September 1994.*

There are two special features about India's milk production. First, India has achieved its production almost entirely by using farm by-products and crop residues as cattle feed, unlike some developed countries like the USA, that divert grains from human to cattle feed. One estimate states that it takes roughly 7 kg of grain in feed lots to produce 1 kg gain in the live weight of animals. Second, the milk production in India has, so far, remained largely with small farmers and marketing strengthened by cooperatives run by them. Apex cooperative milk producers' federations were also set up in Punjab and Haryana in the mid-1970s. There was similar expansion in the poultry sector and egg production increased significantly in Punjab and Haryana. For millions of farmers, the integration of milk production with the crop production system is an important source of supplementary income. The Economic Survey for 2002–3, by the Government of India, points out that in 2001–2 India produced 84.6 million tonnes of milk, and that the contribution of milk alone to the country's GDP was higher than that of paddy, wheat or sugarcane.[6]

WATER SUPPLY

Most irrigation and multi-purpose projects executed in India in the past had invariably considered the drinking water requirements en route as a basic need to be provided for but had not given this benefit a specific monetary value. In many cases, drinking water is not even mentioned as one of the main benefits of the project, as happened in the case of the BNP.

The integrated river system of the Ravi-Beas-Sutlej, supported by the Bhakra and Beas dams and their canals, serves as a major source of domestic, municipal, and industrial water supply to a vast area. The BNP covers the states of Haryana and Punjab, as also the Union Territory of Chandigarh and the National Capital Territory of Delhi. Similarly, the part of Rajasthan that lies in the BNP command also receives such benefits. Groundwater is another major source of water supply but canal irrigation recharges the water table. Till recently, free or highly subsidized electric power helped in tapping groundwater. Most villages along the canal system take advantage of the available water supply for various uses, in addition to irrigation.

Statistics compiled by the Planning Commission show that on an all-India basis, 81 per cent of urban households, and 55 per cent of rural households had access to safe drinking water by 1991.[7] State-wise figures reveal that Punjab and Haryana, along with Chandigarh and Delhi, have a much better coverage. About 94 per cent of urban areas in Punjab and Haryana, as well as 92 per cent of Punjab's and 67 per cent of Haryana's rural households, had access to safe drinking water in 1991.

Most Indian metropolitan areas are now dependent on long distance transfer of water for meeting their increasing demands. Delhi, situated on the banks of the Yamuna river, is unable to meet all its raw water demands with the supplies of the Yamuna, particularly in the non-monsoon months. The Governments of Haryana and Uttar Pradesh divert the entire low season flows into their canal systems at Tajewala. In 1994, the chief ministers concerned had arrived at an agreement on sharing the flows of the Yamuna. This allocation will be feasible only after storages are built upstream. Till then Delhi will need other sources for meeting all its demands.

---

[6] Government of India, Ministry of Finance, Economic Division 2003—Economic Survey 2002–3, para 8.3.

[7] The Planning Commission, *Tenth Five Year Plan (2002–7)*, vol. III, p. 81.

The city of Delhi, which had 9.4 million people in 1991, already crossed 13 million people in 2001, and is still growing. The Delhi Master Plan projects the likely population in 2021 as 23 million. In addition to tapping ground-water and its traditional source of the Yamuna, Delhi is drawing waters from the Bhakra reservoir on the Sutlej, as also the Ramganga dam in the Ganga system. It plans to draw additional water supplies from the Tehri, Renuka, Lakhwar-Vyasi, and Kishau dams for meeting its future requirements.

The present water supply is around 650 MGD, or 3000 million litres per day, which means a per capita gross availability of 274 litres per day. However, large leakages, unaccounted for water and other ills of gover-nance, have left the city thirsting for water. Delhi is, perhaps, the only state in India where water supply is given to all its 219 rural villages, 126 urban villages, 44 resettlement colonies and over 1200 slum clusters. The Delhi Jal (Water) Board has projected that it may not be possible to meet the minimum needs of the city unless many concurrent steps such as locating new sources, conservation and management of available water, upgrading and rationalizing the water distribution system, waste reduction, and re-use of piped water are vigorously pursued.

What is relevant in the context of the present study, is that the Bhakra waters have been meeting a sizeable part of the water needs of Delhi for many decades now. The stored waters from Bhakra are released in accor-dance with past agreements, carried through the Bhakra main canal, and conveyed to Delhi through the Narwana branch and the Western Jamuna canal. The Haiderpur and Nangloi water treatment plants at Delhi are dependent on the raw water supplies from Bhakra. The release of water from the BNP for the Delhi water supply is monitored at high levels. For instance, in May 2000, the Supreme Court of India ordered that 125 cusec of water be released ex-Nangal for the water treatment plant at Nangloi, in addition to whatever was being released earlier. The CWC was directed to monitor this release and receipt of water.

In the entire north-western region of India, Ludhiana was, hitherto, the only city with over 1 million people. Amritsar has now emerged a second in 2001.Though it is situated close to the Sutlej river, Ludhiana's water supply is based on local sources, which have been found to be unsatisfac-tory. There are a number of cities with populations between 100,000 and 1 million in this region. These include the towns Jalandhar, Patiala, Bathinda, and Hoshiarpur in Punjab, and Panipat, Yamunanagar, Rohtak, Hisar, Gurgaon, Karnal, Sirsa, Jind, and Faridabad in Haryana. Not all of them are in the BNP area. However, those that are, stand to gain by sourcing their water to the BNP system.

The populations of Haryana and Punjab, as per the 2001 census, are 21 million and 24 million respectively. Chandigarh supports another 0.9 million people. The rural population is around 71 per cent in Haryana and about 66 per cent in Punjab. A significant part of this number who live in the BNP canal command are the beneficiaries of the waters in the canal system, wherever the local groundwater source does not serve this purpose. The Government of Haryana claims that all the identified problem villages of Haryana have been provided with safe drinking water supply by 1992. Punjab claims that 83 per cent of 'scarcity villages' have been covered by water supply schemes. The availability of a dependable source of potable water for millions in the project area and its neighbourhood translates into better health for the people.

India's National Water Policy (2002) accords the highest priority to drinking water among all its possible uses. It states that adequate safe drinking water facilities should be provided to the entire population both in urban and rural areas. It directs that irrigation and multi-purpose projects should invariably include a drinking water component, wherever there is no alternative source of drinking water. Drinking water needs of human beings and animals should be the first charge on available water. Though the BNP was approved five decades ago, the mandate of the National Water Policy was anticipated and has received full consideration in it.

INDUSTRIAL DEVELOPMENT

The BNP command is home to some of the largest and well-known industries of India. After Partition, this area attracted a large number of displaced persons from Pakistan who had a background of trade and industry. They soon developed industrial enterprises such as manufacturing bicycles, sewing machines, diesel engines, steel fabrication, and textiles. Many large industrial complexes have developed since then.

Ludhiana is called the 'Manchester of India' or the hosiery hub of India and rightly so, for it has proved its industrial mettle and strength, be it bicycles or allied components, knitting yarn, woollen garments, sewing machines, or other industrial products. Hero Cycles Ltd, for instance, is considered the largest manufacturer of bicycles in the world. A recent study made by the Confederation of Indian Industries (CII) has identified three cities in the BNP region, namely Chandigarh, Ludhiana and Amritsar, as among the country's hottest business destinations.[8]

[8] Study by Bibek Debroy for the CII. See report in the *Indian Express*, 21 April 2003.

It is not merely big or medium industries that mark this region. There are a large number of small-scale industries in this area as well. For instance, Ludhiana has over 22,000 small-scale units and about sixty large and medium ones. Punjab had only around 8000 small-scale industries in 1966–7. This has now grown to around 200,000, employing nearly a million people. Haryana's story is similar. It should be noted that small industries tend to be more labour intensive.

With the abundant production of raw materials such as rice, wheat, sugarcane, cotton and oilseeds, and the electrification of towns and villages, a fillip was given to various industries, particularly small and cottage industries. Heavy industries also developed with the availability of abundant and cheap power. The Nangal fertilizers factory was among the earliest of these.

A significant trend noted in Punjab is the conscious effort made towards 'ruralizing industry'. Growth has taken place to a large exent in the rural areas. Many small and large industries moved to rural locations, with the benefit of increased employment of people drawn from these areas. Unfortunately, data in this regard are not easy to obtain. However, the studies that exist relating to the impact of industries migrating to rural areas show greater employment opportunities for villagers. Educated and/or trained women have gained the most in this respect.[9]

RURAL ELECTRIFICATION

Prior to the BNP, the position of power in Punjab and Rajasthan was fairly rudimentary and electrification, if any, was confined to the cities alone. In 1951, only 42 out of 11,947 villages of Punjab had electricity. Small generating units located in a few towns catered to a limited urban population. However, the abundant electricity generated by the project led to a dramatic change in the situation, and set the pace for rapid electrification/development. Availability of power induced the farmers to opt for electricity-operated tube-wells. Entrepreneurs were enthused to set up new industries.

Punjab was the first state in India to have provided electrification to *every* village and rural community by 1975–6. In a similar manner, Haryana, after it became a separate state, also pursued rural electrification programmes and declared 100 per cent rural electrification by 1975–6. In 1991, Punjab had 12,428 villages, all of which were electrified. In the same year, Haryana had

[9] Dev, D.S., *et al.*, *Rural Industrialisation, Lessons from a Case Study*, Paper in the seminar at Punjab Agricultural University, Ludhiana, May 1993.

6759 villages, all of which were electrified. In Himachal Pradesh, there were 16,997 villages in 1991, of which 16,819, that is, 99 per cent, were electrified. In Rajasthan, in 1991, there were 37,889 villages. Of these, 34,937 were electrified in 1991.The National Capital Territory of Delhi has 199 villages, which are 100 per cent electrified.

The provision of electrification everywhere had vastly helped the development of groundwater potential, as well as the setting up of small and cottage industries. The project authorities indicate that 128 towns and 13,000 villages were electrified by the BNP.

PISCICULTURE

The project report does not speak of the fisheries, if any, that existed in those days. The Bhakra dam authorities consider that prior to the construction of the dam not much fish existed there. However, after the creation of the Bhakra reservoir, fish breeding began and is now thriving. The Central Inland Fisheries Research Institute has recorded some fifty-one species of fish from the Gobind Sagar reservoir.

The following are the dominant varieties of fish being encountered in the catches of the reservoir:
- *Labeo rohita*
- *Catla catla*
- *Cirrhina mirgala*
- *Labeo calbasu*
- *Cyprinus carpio*
- *Hyothalmichthys molitrix*
- *Tor putitora/Labeo dero*
- *Mystus seenghala*
- *Ctenopharyngodon idellus*
- *Labeo bata.*

There is a considerable variation in fish catches during the different months of the year. However, on an average, around 3 tonnes of fisheries, that is about 1100 tonnes of fishery products, are raised and sold each year. In 2000–1 the Fisheries Department realized a revenue of nearly Rs 4 million by way of royalty, license fee and so on. A total of 1400 fishermen are engaged, full-time, in fishing activities in the reservoir. The Bhakra reservoir is, thus, an important source for raising fisheries in the region.

As the reservoir lies within the state of Himachal Pradesh, the Himachal Pradesh Government is actively engaged in further developing the fisheries in Gobind Sagar.

CAPACITY BUILDING

The Bhakra dam was among the earliest projects that independent India built. An important policy decision taken in 1950 helped develop capacity building with respect to water resources development projects. There was a debate in the country on the pros and cons of undertaking the construction of the Bhakra project, either as a departmental one through a government agency, or as work to be let out on contract to a construction agency in the private sector. Many foreign firms were interested in undertaking the work and some German and US firms were under consideration for the Bhakra project. However, the Government of India decided on departmental construction. This gave Indian engineers, planners, and workers the unique opportunity of training themselves in constructing river valley projects. The construction of the Bhakra-Nangal project thus became well known for imparting training to engineers, technicians, and unskilled labourers. With the construction of the first batch of river valley projects like Bhakra-Nangal, Hirakud, Damodar Valley, Rihand, and Gandhi Sagar, engineers based in different parts of the country demonstrated their ability to rise to the occasion and showed that they were, indeed, second to none in designing and constructing large projects. The CWC and its engineers also benefitted with experience gained from interactions with the engineers of this project. A large number of workmen also received training in the process.

TOURISM AND RECREATION

The large Gobind Sagar lake that was formed upstream of the Bhakra dam has become a major tourist attraction. Annually, about 300,000 people visit Bhakra. The Krishna Raja Sagar lake, with its Brindavan garden in Mysore, was a 'pace setter' in India, in developing tourism as an integral part of a river valley project. A large number of people not only from Punjab and Haryana but from all corners of India flocked to Bhakra to view the engineering marvel, as well as the bounties of nature that have been developed there. The existence of religious centres like the Anandpur and Kiratpur Sahib gurudwaras, as also the Naina Devi temple in the vicinity of the project area, and the architectural attraction of the nearby Chandigarh city planned by Le Corbusier, have made a visit to Bhakra and nearby places an attractive tourism package. The lake is important for its migratory birds that come from Siberia. The green forests surrounding the lake, the high hills all around, the fresh air and calmness provide an ideal environment.

The Himachal Pradesh Government has constructed a huge water sports complex at Bilaspur on the banks of the Gobind Sagar reservoir. Training is imparted here in swimming, surfing, water skiing, rowing, canoeing and so on.

Recently the Government of Haryana has initiated farm tourism, in which interested tourists are taken to typical old-style villages and their surroundings. This involves staying in farmhouses and visiting cultivated fields. Such tourism could well be combined with visits to the project and its components like canals and structures located nearby.

In the recent decade or two, the tourism aspect had to be moderated due to new security concerns that had arisen in the area. The proximity of Bhakra to the border and the belligerent militant movements that were witnessed in the region called for restrictions in providing access to certain vital installations and areas. Notwithstanding this factor that was necessary for national security, Bhakra-Nangal still continues to be an important recreation and tourism centre.

TOWN AND COUNTRY PLANNING

It was noted earlier, in the context of the resettlement of the refugees after the Partition of India, that new canal colonies were developed in the Bhakra-Nangal region. From past experience gained in united Punjab, in incorporating town and country-planning aspects as parts of the irrigation canal colonies, Bhakra was a forerunner in independent India. A number of attractive features have been incorporated in the development of the villages and cities in the BNP region. The experience gained in the planning and construction of the new capital, Chandigarh, also gave an impetus to the incorporation of many desired improvements relating to country planning while also developing new irrigation colonies in the area. Many villages in the project region were considered as models of country planning, integrated with the development of irrigated agriculture.

EMPLOYMENT GENERATION

During the days of the construction, about 12,000–15,000 persons were engaged in various activities connected with the project, for about 15 years. Not only qualified engineers but also ordinary workers acquired construction skills while working in the BNP. This enabled them to obtain gainful employment and better wages in similar activities elsewhere.

A criticism that was voiced in the days when construction on the scheme started was that the employment opportunities were lower in the BNP because of the use of mechanized equipment. This proved to be a rather narrow way of looking at the picture. The manner of construction and the extent of mechanization were dictated by the type of the dam and the difficult site conditions. What has not been realized or highlighted, is that the project enabled millions of people to obtain gainful employment in agricultural and related activities for decades after the completion of construction.

ADVANTAGE FOR WOMEN

The lives of women are closely linked with water. Any traditional picture of a semi-arid region like Rajasthan and parts of Haryana and Punjab will show women and girls balancing pots of water on their heads and in their hands, and trudging miles and miles to provide the family with the requisite water needed for the household. The BNP and its vast canal system made a drastic change in this traditional picture. Its 1100 km of main canal and branches, and 3368 km of distributaries and minors not only provided irrigation facilities to new areas but also brought water closer to villages and homes in the irrigation command area. Abundant water sources were brought closer to homes by the BNP. Women were the greatest beneficiaries of this. Their drudgery, labour, and time lost in procuring water was vastly cut down. If suitable allied policy and institutional measures are pursued, this saved time could be diverted towards their better education and acquiring skills. The general improvement in health by improved water supply and sanitation should also be acknowledged. With the full development of irrigation and particularly after the spread of paddy cultivation, the development of tube-wells also occurred. This enabled the exploitation of groundwater sources, in addition to canals for water.

The cultural setting of Indian tradition, unfortunately, has not given women equality of rights with men. As noted earlier, mere provisions in the Indian Constitution or the laws of the land will not give them their due status. A drastic change in the mindset is needed. This negative aspect of Indian society has been noted by many commentators in the context of women not receiving their due in irrigated-agriculture, even after the Green Revolution. The development of the canal system has bought water close to the homes, thus relieving women from their drudgery of carrying it from far-off places. However, this positive aspect and the resultant benefit to women has hardly been noticed or recognized.

Indirect Benefits

'MULTIPLIER EFFECT' OF THE PROJECT ON THE ECONOMY

Many primary and incidental benefits from the BNP have been noted earlier. There are, in addition, many indirect and induced benefits and consequences not discussed in the above paragraphs. These direct consequences, in turn, generate a number of indirect and induced effects. These are those that arise out of the myriad linkages, both backward and forward, between the direct consequences of the project and all other sectors of the economy. For instance, hydropower from the dam not only provides electricity to urban and rural regions but also enables increased outputs of industrial commodities (for example, fertilizers, chemicals, and machinery). This, in turn, requires inputs from other sectors of the economy (such as steel). Increased agricultural outputs encourage the setting-up of food processing industries. There are, thus, many backward and forward linkages.

The WCD Report recognizes the importance of indirect impacts. It says:

> As with livelihood enhancement, the broader impacts of irrigation projects on rural and regional development were often not quantified. Dams, along with other economic investments, generate indirect economic benefits as expenditure on the project and income derived from it lead to added expenditure and income in the local or regional economy. The WCD case studies give examples of these 'multiplier' benefits resulting from irrigation projects....
> ...a simple accounting for the direct benefits provided by large dams—the provision of irrigation water, electricity, municipal and industrial water supply, and flood control—often fails to capture the full set of social benefits associated with these services. It also misses a set of ancillary benefits and indirect economic (or multiplier) benefits of dam projects'.[10]

It is, however, another matter that having stressed the great importance of taking into account the multiplier effect, the WCD Report failed to give due recognition to such positive impacts of large dams.

The economic development of Punjab and Haryana in the last three to four decades reveals that a very close relationship exists between water resources development and the structure of the economy. The linkages induced by input demand and output supply generate growth impulses which are transmitted from one sector to another of the economy. Additional employment opportunities are generated in the non-farm sector, for instance, for servicing and repairing of machinery and implements.

Their effects extend to the whole project region, as well as outside it. In other words, the major outputs of a dam project will have inter-industry linkages that result in increased demands for output in other sectors of the economy. Increased agricultural and industrial output leads to the generation of additional wages/income for households, in turn leading to the higher

---

[10] WCD Report, 2000, pp. 100 and 129.

consumption of goods and services. All these lead to changes in output-input of various such sectors. These could be estimated and aggregated in terms of the 'multiplier value', reflecting the ratio of the total direct and indirect impacts of the project to its direct impacts alone. There are huge employment opportunities created for the migrant labourers who came to the BNP area from other regions on a regular basis (largely from Bihar and Uttar Pradesh). They have the possibility of not only earning high wages but also of acquiring agricultural and mechanical skills during their work in the BNP area. Similarly, the 'food for work' programme creates employment opportunities in various parts of India.

Increase in personal disposal income will also create demand for services like transport, communication, and recreation. The increasing preference of consumer durables like scooters, cars, refrigerators, televisions, and kitchen gadgets, will, in turn, need repair and maintenance services. Such an overall change in the socio- economic scene and its beneficial effect on the tertiary sector have been witnessed in Punjab and Haryana.

Analytical frameworks for such estimations of regional value-added multipliers are available.[11] These tools, which are essentially in the nature of multisector models, include: (a) Input/Output models; (b) Social Accounting Matrices (SAM)-based models; and (c) Computable General Equilibrium (CGE) models. The particular choice to be made in a specific case depends on the mechanisms through which the impact is transmitted in the region of interest and the database available.

Such studies on the indirect consequences of dams, sponsored by the World Bank, were made by Ramesh Bhatia and R.P.S. Malik with respect to the BNP, using SAM-based models, under 'with' and 'without' project conditions. These indicate that the multiplier may lie in the range of 1.8 to 1.9; which meant that for every rupee of additional direct output of the project, another 0.8 to 0.9 was generated in the form of downstream on indirect benefits.[12]

The figure would have been higher than 2.0 if the full accounting of industrial development, significant additions to the agricultural labour pool

---

[11] For instance, see the pioneering work of Peter Hazell and R. Slade (1982) on the Muda dam project in Malaysia, or that by P. Hazell and C. Ramaswamy on the impact of the Green Revolution in south India (1991). Johns Hopkins University Press, Baltimore, Maryland.

[12] Unpublished reports on the studies by Ramesh Bhatia and R.P.S. Malik, Draft Report, March 2003, and Ramesh Bhatia, *et al.* (eds), Indirect economic impacts of dams, under publication by the World Bank.

and remittances of the migrant seasonal labour to other regions were included. It is an interesting fact that in the last decade, the number of migrant labourers working in the states of Punjab, Haryana, and Rajasthan was approaching a million.

MIGRANT LABOUR

The BNP has created huge employment opportunities in agriculture and related sectors. Plentiful employment opportunities available on a continuing basis for hired labour every year lures hundreds of thousands of labourers from far-off poor regions of the country—like Bihar and Uttar Pradesh, where wage rates are low and unemployment is very high—to migrate to this region in search of employment and better wages. Some of them have settled down permanently in the project region itself, while others migrate every year.

Ramesh Bhatia cites a study carried out recently by the Punjab Agricultural University, wherein it has been estimated that during the lean period of agricultural operations, the number of migrant labourers employed in Punjab was 387,000 and this number increased to about 774,000 in the peak period (Sidhu *et al.*1997). About 93 per cent of the migrants belonged to the poor states of Bihar and Uttar Pradesh. It has further been estimated that the number of migrant labourers has increased by over 35 per cent in 1995–6 as compared to 1983–4.

The wage rates which these migrant labourers received in their native villages were very low as compared to what they got in Punjab. It was stated that about 46 per cent of the migrant labourers were getting less than Rs 300 per month for being employed on a permanent basis in their native villages, while in Punjab they were getting almost 200 per cent higher wages.

Ramesh Bhatia estimates in his studies that the earnings of the migrant labour force engaged in crop production in Punjab and Haryana during 1995–6 was Rs 5344 million (US$ 114 million). Out of their earnings, they remitted Rs 3548 million (US$ 75 million or 66 per cent) back to their native places. About 18 per cent of the migrant labour utilized their savings for creation of assets in their native places.

The remittance of such huge amounts of money to relatively far-flung poor regions from where the migrant labourers come, has helped improve the living conditions of people living there, and has created its own further downstream effects. Most migrant labourers utilized their savings for the purchase of consumer durables such as televisions, radios, bicycles, sewing machines and so on. A good portion of them also utilized their savings from

Punjab for the repair and construction of their houses in their native villages. Some used the savings to either lease land for increasing their operational holdings, or for buying new pieces of agricultural land. Parts of the savings were, of course, used for improving daily consumption.

Migration of people in search of employment has been rising since the 1990s. There are no official figures of these numbers. Even the Census and National Sample Survey ignore or underestimate short-term migration owing to their procedural and definitional problems. The number, however, runs in millions. This is a phenomenon which has eluded quantification for essentially political reasons. P. Sainath, in a newspaper article which appeared in the *Hindu* of 15 March 2004, complained that millions of Indians who really wanted to vote in the national elections that were being held would not do so because of the timing of the election in April-May.[13] He said that it was precisely in these months that there was a large exodus of migrant labour (see box item—the millions who cannot vote).

Excerpts from the *Hindu*
THE MILLIONS WHO CANNOT VOTE
15 March 2004

P. Sainath

By having elections in April when millions of the poor
migrate in search of work, we are simply excluding
an ever-growing number of citizens from the vote.

Millions of Indians who really want to vote in this election will not. The rural poor, far more than the chattering classes, are the pillars of our electoral system. The vote is the one instrument of democracy they get to exercise. And they do that with telling effect, often using it to go out and change governments. This time, many of them cannot vote. The timing of the polls ensures that. It is in April–May that quite a few distressed regions see their largest exodus of migrant labour.

....

....

In late April, there will be Biharis still in Punjab or Assam, Oriyas in Andhra Pradesh, Gujarat or Chhattisgarh. ....

In April–May there will be countless millions of them. Forced to scrape out a living away from home. ....

By denying them that, we undermine them, ourselves, and democracy too.

---

[13] Sainath, P., the *Hindu*, 15 March 2004.

12

SEDIMENTATION IN THE BHAKRA RESERVOIR

Reservoir Sedimentation

Sedimentation of reservoirs is an unavoidable and natural process that sets in simultaneously with the commencement of storage. All reservoirs formed by dams on natural rivers are subject to some degree of sedimentation. However, this issue is of vital concern to water resources development projects. If it is possible to correctly forecast the amount of sediment inflow, as also the pattern of deposition, and if the project design takes care of these, such an arrangement should be able to reasonably ensure unimpaired performance during its expected useful life. The natural processes involved in erosion, entrainment, transportation, deposition and compaction of sediment, as also their effect on the useful life of reservoirs, are, however extremely complex. Human activities during the post-project period extending over decades, if not a century or more, like deforestation, excavation, quarrying, mining, construction, etc., in the catchment area further contribute to the problems. It would be a difficult task to correctly foresee them.

The issue of importance to the water resources project planner is to estimate, as realistically as possible, the likely rate of sedimentation and the period of time by which this sedimentation will interfere with the effective functioning of the reservoir project. The planner is, thereafter, required to incorporate requisite provisions for sediment storage in the reservoir, so that it will not impair the project functions during its 'designed life'. Sedimentation and reduction in capacity is normally a gradual process. Reservoirs do not have a defined life in the strict sense, in as much as it is not a question of 'ON' or 'OFF'. They belong to the category of systems that gradually degrade in performance without any sudden 'non-functional' stage. Human activities in the post-commissioning period in the catchment area and reservoir should also be carried out with due care and a sense of responsibility.

After the reservoir is formed and the scheme gets commissioned, in the first phase, it will continue to deliver the full planned benefits. Thereafter, there will be progressively reducing benefits but the scheme will still be economically beneficial, even with the reduced benefits. In the third phase, the sedimentation could cause difficulties in operation. Soon thereafter, the difficulties become so serious that operations becomes impossible, or it is no longer beneficial to operate the scheme.

Sedimentation in the Bhakra Reservoir

In the case of the Bhakra reservoir, sedimentation studies were made in 1947–8. These used the suspended silt load observations made for the Sutlej river during the period 1916–39. They indicated that the average annual suspended sediment load was 34.5 million tons. The bed load carried by the river was not measured, and was taken as 15 per cent of the suspended load, thus aggregating to 40 million tons (40.64 million tonnes) per annum. For working out the likely volume that would be occupied by the deposits, an estimate of the density of these deposits was made after considering the experience in the Elephant Butte and Boulder dams of the USA. It was assumed that all coarse sediment would be deposited in the head reach of the reservoir, forming a delta below full reservoir level, and that its gradual progress towards the dam would also lead to encroachment in the live storage, too.

The annual sediment was estimated at 2417.7 Ha.m or 19600 acre feet equivalent to 4.28 Ha.m/100 sq. km/ year (or 90 acre feet/100 sq. miles/ year). It was also estimated that full silting would take place in 535 years. Irrigation benefits were expected to be affected after 25 per cent silting of the live storage, which was calculated to be after 135 years. In 286 years, 50 per cent silting of live storage was anticipated. Notwithstanding these, all financial forecasts were worked out on the basis that the economic life was only a hundred years.

In the earlier years of India's planning of river valley projects, the concept of the life of reservoirs was vague. The estimation of useful life was done in a varied manner from project to project, and from period to period, with different assumptions.

The phenomenon of sedimentation and the concept of the life of a reservoir have been studied in depth over many decades now. Provision for sediment deposition in reservoirs planned in India after Independence was made on the basis of experience and data available from abroad. Dr A.N. Khosla, who was responsible for the investigation and planning of the Bhakra dam and who later became the first Chairman of the Central Waterways Irrigation Navigation Commission (the parent body from which today's CWC and Central Electricity Authority [CEA] have emerged), did some pioneering studies on this subject. He reviewed the position of reservoir sedimentation in the 1950s based on the data of some 200 reservoirs from all over the world including the USA, Africa, and China, and developed some enveloping curves for the annual sedimentation rate for major and minor catchments. He concluded that the sediment rate for major catchments varies from 3.57 to 4.76 Ha.m/100 sq. km/year and for minor catchments from 3.8 to 12.8 Ha.m./100 sq. km./year. Until 1965, his recommendations were the guiding factors in the design of reservoirs in respect of sedimentation, and the life of the reservoir was calculated as the period required to fill the dead storage capacity.

After the mid-1970s the CWC began insisting that sediment inflow rates should be based on reservoir survey data. By the mid-1970s it was considered that the '50 year sedimented position' of the reservoir should be used in the simulation of working tables for projects.

The Indian Standards (IS) 12182 (1987) Guidelines for determination of the effects of sedimentation on the planning and performance of reservoirs has the following main features:

- The sediment rate is to be determined on the basis of observations of river sediment and reservoir surveys.
- Methodologies for trapping efficiency and sediment distribution are specified.
- The live storage is to be planned such that the benefits do not get reduced on account of sedimentation for a period of 50 years for irrigation or 25 years for hydropower.
- The outlet levels are to be planned such that sedimentation beyond the outlet, likely to cause operational problems, would not occur for 100 years.
- Procedures for simulation studies are specified.

Systematic reservoir surveys were undertaken in India since 1958 under a coordinated scheme. Data in respect of some 144 reservoirs are now available to the CWC. A systematic analysis and study of the capacity survey data has also been done by CWC. It was observed that in reservoirs with a small sluicing capacity with respect to normal floods and no reservoirs above them, the siltation rate is comparatively high in the first 15–20 years and, thereafter, falls off. The surveys revealed that the sedimentation rates in some reservoirs are higher than those envisaged at the planning stage. They also confirmed that silting takes place both within dead storage and in the live storage of the reservoirs. The CWC study showed that the rate of sedimentation is the maximum in the Himalayan region (Indus, and Ganga-Brahmaputra basins) as compared to the rest of India. Of 144 reservoirs for which data on reservoir surveys are available to the CWC, Bhakra is the only case where twenty-seven repeated surveys were made in a period of 42 years (1958 to 2000). Next came the Pong reservoir with fifteen surveys in 26 years of operation (1974 to 2000). Other than these two, there are no cases with more than ten repeated surveys.

Data relating to the Bhakra reservoir are summarized in Table 12.1. This clearly reveals that the actual rate of silting has remained nearly constant at around 6 ha.m /100 sq. km/year, as against the assumed rate of 4.29 ha.m/ 100 sq. m/year. Even at the present higher rate of silting in the reservoir, the project will fully serve its originally assumed designed economic life of a hundred years and more.

The cumulative sediment deposited in the reservoir between 1958 and 2000 was 1482.24 MCM. Unlike the earlier assumptions, silting has occurred both in the live and dead storages. The details are given in Table 12.2. It is observed that the total loss of capacity in 42 years of its life is around 15 per cent. The silting in the live storage is 10 per cent.

The repeated reservoir surveys indicate that the average annual rate of siltation is 35.29 million cubic metres (28,612 acre ft). However, as the dam construction works continued till 1964, the Bhakra Dam Organization worked out the average rate of siltation in the reservoir for the period 1965 to 2000, ignoring the initial period up to 1965. This works out to 34.62 million cubic metres (28,067 acre ft). Thus, the actual rate of siltation is not very different from the design figure of 33.61 million cubic metres (27,250 acre ft).

The project has started observing, at selected locations in the reservoir, the *in situ* density of the sediment deposit. This is with a view to determine the changes in the sediment density with time. The density of silt adopted for design purposes was 77.5 lbs/c.ft (1.24 gm/cc). So far, the actual average density is reported to vary from 86 to 95 lbs/cft (1.38 to 1.52 gm/cc). If the deposits get densely consolidated over the years, this would result in the increased useful life of the reservoir.

The 1951 project report stated that the life of the reservoir was likely to be over 500 years, but was taken as 100 years only for the purpose of working out financial returns and projections. On the basis of the latest actual observations and projections therefrom, the loss of dead storage capacity will take 138 years and the live storage in 362 years. Thus, while the actual rate of siltation is higher than the design assumption, the conservatively adopted useful life of over a hundred years continues to be more

Table 12.1: Sedimentation in the Bhakra Reservoir

River : Sutlej
Catchment area : 56,980 sq. km
Assumed rate of silting : 4.29 Ha.m/100 sq.km/year

S. no.	Year of survey	Years since first impoundment	Reservoir capacity M.cu.m	Total loss of capacity M.cu.m	% loss of original capacity	Average loss of capacity per year M.cu.m	Rate of silting Ha.m/ 100 sq.km/year
1	1958	–	9868.00	First year of impoundment			0.03
2	1959	1	9867.84	0.16	Neg.	0.16	5.04
3	1963	5	9724.53	143.47	1.45	28.694	5.92
4	1965	7	9631.92	236.08	2.39	33.726	5.71
5	1966	8	9607.71	260.29	2.64	32.536	6.08
6	1967	9	9556.01	311.99	3.16	34.666	6.05
7	1968	10	9523.55	344.45	3.49	34.445	5.76
8	1969	11	9507.11	360.89	3.66	32.808	5.93
9	1970	12	9462.41	405.59	4.11	33.799	5.68
10	1971	13	9447.57	420.43	4.28	32.341	5.65
11	1972	14	9417.53	450.47	4.57	32.176	5.53
12	1973	15	9395.39	472.61	4.79	31.507	5.77
13	1974	16	9341.83	526.17	5.33	32.886	5.65
14	1975	17	9320.98	547.02	5.54	32.179	5.68
15	1976	18	9285.25	582.75	5.91	32.375	

(contd)

Table 12.1: contd

S. no.	Year of survey	Years since first impoundment	Reservoir capacity M.cu.m	Total loss of capacity M.cu.m	% loss of original capacity	Average loss of capacity per year M.cu.m	Rate of silting Ha.m/ 100sq.km/year
16	1977	19	9259.82	608.18	6.16	32.009	5.62
17	1979	21	9160.32	707.68	7.17	33.699	5.91
18	1981	23	9091.14	771.86	7.87	33.776	5.93
19	1983	25	9034.62	833.38	8.45	33.335	5.85
20	1984	26	9007.28	860.72	8.77	33.105	5.81
21	1986	28	8932.94	935.06	9.48	33.395	5.86
22	1988	30	8855.72	1012.28	10.26	33.743	5.92
23	1990	32	8736.35	1131.75	11.47	35.367	6.21
24	1992	34	8681.57	1186.43	12.02	34.095	6.12
25	1994	36	8585.29	1282.71	13.00	35.631	6.25
26	1996	38	8533.49	1334.51	13.52	35.187	6.16
27	1998	40	8477.84	1390.36	14.09	34.759	6.10
28	2000	42	8385.62	1482.24	15.02	35.29	6.20

Sources: 1. Government of India, Central Water Commission—Compendium on Silting of Reservoirs in India, 2001.
2. Sedimentation Studies Bhakra Reservoir. BBMB—June 2001.

than realistic. If upstream catchment area treatments are pursued, and additional storages come in the upstream, the life of the reservoir would be more than the present projections.

The Baspa Hydroelectric Project (HEP) on the Baspa tributary, Nathpa-Jhakri HEP just downstream of Karcham, and Sanjay Vidyut Pariyojana, a right bank tributary of the Sutlej, are the important projects that have come up upstream of the Bhakra dam. All these schemes include provisions for catchment area treatments. These should tend to reduce the sediment inflow into the Bhakra in the future.

Trap Efficiency of Reservoir

The rate of silting in a reservoir depends, in addition to sediment production in the catchment, on many other factors. They include the trap efficiency of the reservoir, ratio of reservoir capacity to the run-off, gradation of silt, manner of the reservoir operations and so on. In the case of a storage reservoir with a larger capacity in relation to the inflow (say if the capacity is more than the annual inflow), it would hardly spill over the dam and, hence, the trap efficiency would be very high. Similarly, in a reservoir with very low capacity but much larger stream flows, the trap efficiency will be low. The Bhakra reservoir has a high trap efficiency. Its actual average trap efficiency is 99.4 per cent, as against the project assumption of 85.5 per cent. The assumption made earlier that more silt would flow downstream with density currents has not yet materialized.

Silting in Live Storage

It has been indicated earlier that, contrary to earlier conventional assumptions in the 1950s and 1960s, silting in the live storage region is a confirmed fact. This will have its impact on the life of the reservoir. This was the picture seen from the earliest reservoir surveys in Bhakra. In the case of the Bhakra (Gobind Sagar) reservoir, silting in the dead and live storages are given in Table 12.2.

Mitigation Measures

The usual measures undertaken to mitigate the reservoir sedimentation problem are:

PREVENTIVE MEASURES

1. Watershed management activities in the catchment area
2. Creation of off-channel auxiliary/retarding reservoirs

Table 12.2: Siltation in the Bhakra Reservoir

	Storage in million cubic metres		
	Dead storage	Live storage	Total
Original capacity	2431.81	7436.03	9867.84
Silting 1958–2000	761.22	721.02	1482.24
Present reduced capacity	1670.59	6715.01	8385.60
Percent loss in capacity	31.30	9.70	15.02

Source: BBMB.

3. Creation of bypass canals to divert sediment-laden flows around the reservoir

REMOVAL OF SILT

1. Sluicing/flushing sediments by optimized reservoir operation
2. Removal of silt by mechanical means

Off-site disposal of sediment is very expensive, needs a vast disposal area and could create environmental problems. Sediment removal by mechanical means is also neither suitable nor cost-effective. Watershed management activities are effective though it takes a long time for implementation. In fact, this approach was adopted in the case of Bhakra by a centrally sponsored scheme for soil conservation under the Ministry of Agriculture, known as the RVP Sutlej scheme, initiated in 1962–3. An area of 19,830 sq. km, which forms the entire catchment of the Sutlej within India, was covered by this scheme. The prioritization of the area for treatment is based on the recommendations of the All India Soil and Land Use Survey Organization. There have been continuous and heavy inroads into the forest areas of the catchment. In addition, many developmental activities like road construction serve as negative factors, overshadowing the beneficial impacts of the watershed development works undertaken. There is a need for a holistic approach to the treatment of catchment areas with low-cost conservation practices, involving the local community in a big way, and to bring about the integrated development of micro-watersheds. The Government of India, Ministry of Energy constituted a committee in 1981 to examine the siltation problem and suggest afforestation and other measures for the Sutlej (Bhakra) and Beas catchment areas. Accordingly, it recommended a plan for afforestation, fencing, monitoring and evaluation works. Since the Gobind Sagar reservoir is in the Himachal Pradesh territory, the BBMB entrusted the afforestation to the forest department of that state. With the funds provided by the BBMB, a work plan for afforestation and other soil conservation measures has been prepared by the Himachal Pradesh Forest Department.

One of the difficulties in studying the life of reservoirs while a project is planned and approved, or even after several years, is that it would be tricky to predict or anticipate what changes would occur in the catchment area. For instance, the Nathpa-Jhakri and Kol dam projects, now under construction, were not envisaged in 1951. These dams and other storages would act as check dams and decrease the silt inflow into Bhakra. Correspondingly, other types of activities would have an opposite effect.

The conventional methods of conducting reservoir sedimentation surveys are time-consuming and not always highly accurate. In recent decades, modernization in these have been introduced in India. The HYDAC system was adopted under a CWC project in case of the Bhakra, Tungabhadra, Sriramsagar, Hirakud, and Ukai reservoirs. Since the data collection and analysis are fully computerized, this eliminates human error and becomes more reliable.

The remote sensing technique has now emerged and established itself as a useful tool that is cost and time-effective to estimate capacity loss. Satellites provide land surface imageries repetitively, from which water-spread areas of reservoirs can be identified. The satellite imageries for various dates and the corresponding gauge observations enable the development of an area-elevation relationship. This technique can be used for working out reservoir capacities upto minimum draw-down levels. This system has been applied to reservoir sedimentation studies. Thus, the application of remote sensing, GIS, and simulation studies are future powerful tools to estimate reservoir sedimentation.

Over the years, indiscriminate land use, irrational methods of cultivation, deforestation and over-grazing could lead to heavy soil erosion in the catchment area. Large-scale felling of trees for domestic needs and increased industrial demands could also aggravate the problem. It is believed that at least some of these could have led to the greater silting that has been noted. Therefore, serious efforts need to be made to check all these undesirable practices in order to arrest the rate of erosion and reservoir sedimentation.

13

IMPACT OF BHAKRA-NANGAL PROJECT: PEOPLE'S PERCEPTION

General

The BNP was a major river valley project undertaken by independent India, that was later included in the First Five Year Plan (1951–6). The project, in tandem with the Beas complex, forms an integrated system, irrigating over 3.5 million ha of land in Punjab, Haryana, and Rajasthan, generating hydropower and rendering other benefits. Its waters have helped produce enough grains to virtually eliminate famine from the country, and has insulated it from recurring food shortages and dependence on imports. The contribution of Bhakra-Nangal in India's battle against hunger is striking and self-evident.

Planning for water resources development projects and their consideration, as well as approval for execution were standardized after the setting-up of the National Planning Commission. As the BNP was planned, examined, and approved for execution long before that, it would not be correct to compare its various provisions and the manner of its approval and execution with the format settled many years later. However, there is no dearth of commentators and people who do exactly that. They sometimes criticize the BNP and its various provisions, by comparing them with what is being followed in recent years in the case of other river valley projects. For example, when land was acquired for the BNP, it was based on the procedures prevalent in the late 1940s and early 1950s and the laws of the land at that time. Many critics of recent times and even common people have expressed complaints that the terms of their compensation and rehabilitation were lower than the current norms prevailing for other dams in Punjab and elsewhere taken up decades later. Even people who have vastly benefitted by the fruits of the project are swayed by such criticism. This even creates doubts amongst policy makers about the very wisdom of taking up new water resources development projects.

The sphere of influence of the BNP extends over the states of Himachal Pradesh, Haryana, Punjab, Rajasthan, and Delhi. It stabilizes and makes improvements to the old Sirhind canal irrigation system and brings irrigation to vast new areas. It generates significant blocks of hydropower. There are numerous incidental benefits of the project as well. Meticulous data collected by the central and state agencies, with regard to its actual performance in the field of irrigation, power generation and so on over the last four decades and more, show that the project is fully rendering the benefits envisaged in the project report. The previous chapters of this study have made use of these data to draw conclusions about its effectiveness and impacts on various social, economic, and environmental aspects.

The aim of the present study is not only to look at the primary objectives of the BNP but also to examine the secondary and tertiary consequences and its contribution to the development of the region and the lifestyle of the people. Therefore, it was considered that the perception of the people who are benefitted or affected in any manner should also be ascertained by random sample surveys and be made an integral part of the study.

Within the constraints of the limited financial resources, and available time and manpower, field interviews were planned and executed between July 2003 and January 2004 to ascertain the people's views with regard to the various consequences. A basic set of questionnaires was prepared for this purpose. A total of 527 concerned persons were covered by the survey. The people's perceptions as gleaned from the interviews are discussed below.

Displaced Persons due to the Bhakra Reservoir

It has already been stated in Chapter 5 (in the discussion on displacement, resettlement, and rehabilitation) that 375 villages and the old town of Bilaspur were affected, necessitating the shifting of 7209 families from these villages and about 4000 persons from the town. It was also noted that the displaced rural families were offered alternative land within the command area of the Bhakra canals in the Hisar district (which now comprises the Hisar, Sirsa, and Fatehabad districts of Haryana). For this purpose, 5342 ha (13,200 acres) of land were acquired in compact blocks in 30 villages. It was finally noted that 2179 families who opted for resettlement in the irrigation command in the Hisar district were accordingly resettled and rehabilitated. The study sought to examine the perceptions of these rehabilitated people.

URBAN PEOPLE DISPLACED IN THE NEW BILASPUR TOWN

The new Bilaspur township is a fairly well-developed hill town, with its busy marketplace and various other facilities. Interviews with displaced people who were resettled there brought out the following picture:

Most oustees originally had some difficulties in readjusting to their new bearings and starting afresh. The business then was slack and hence, many had to switch over to new types of businesses. They were able to overcome their problems, and in the last few decades, became well settled. They are apparently economically better off now. However, the elderly people have fond memories of their days in the old town. They still recall how they used spring water for drinking, how their old land along the river bank was cultivated soon after the river spill subsided, and so on. There was plenty of free brushwood for them then for cooking purposes. Those who lived along the riverbank benefitted also from the timber that washed down the river from the upstream during a flood. There was nostalgia for their old ways of life and environment. However, they all agreed that the many ills of the present-day life in the new township are similar to those experienced in most other towns in their state or even outside it. They had a general grievance that their sacrifice for the nation had not been fully recognized by the people who were benefitted, or even by the project authorities. Their main grouse was about the lack of educational facilities in the initial phase, say in the first decade or so, which proved to be a setback. Their children were not offered employment by the project authorities even in lower-end jobs. At present, most of the resettled persons are engaged in petty retail businesses.

QUANTUM OF COMPENSATION

The residents were not happy with the amount of cash compensation they received. Their traditional joint family system proved to be a disadvantage for most of them. If they had presented each person or adult as a separate unit of family, they would have got more compensation and more house plots. The cash they received was also spent in meeting the various expenses connected with shifting from the old to the new place, settling down there, and other household expenses. They acknowledged that the government had set up a tribunal for the rectification of all alleged unjust awards but the lack of leadership and education among them resulted in only a very few people approaching the tribunal for redressal of grievances. However, some big landlords were greatly benefitted.

HOUSING

Their houses in the old township were simple structures with some ordinary roofing. At present, most have permanent houses built in higher altitudes of masonry or concrete, with proper roofs which are more durable and weather-resistant.

GENERAL SOCIO-ECONOMIC CONDITIONS

Many stated that the disparity between the rich and poor existed. Most families have one or more members, who are either employed in the army or in other services, as the income from business cannot support the family. After the project, employment opportunities in the tertiary sector have increased. Road transport has enabled the setting up of roadside shops, petrol pumps, small rest houses, vehicle repair shops, and so on. Small-scale tourism has also developed. There are no minor or lift irrigation systems and, hence, their land is hardly being cultivated. They feel that if lift irrigation from the reservoir could be provided, it would benefit them.

The age and health profile of the population indicates that people are generally healthy. A large number of them are senior citizens; some of the interviewees were over 90 years old.

RURAL PEOPLE DISPLACED RESETTLED IN HISAR

Land made available to the 2179 families in the BNP command had to be opened up and made ready for irrigated agriculture as soon as construction work on the project was completed.

Visits to the Hisar/Sirsa/Fatehabad area were made a part of this study, in order to ascertain how the resettled people were living and to obtain their views. The following is the scenario projected by the respondents.

CONDITIONS BEFORE 1960

The area was an arid desert with sand dunes before the advent of irrigation canals. In the initial years, there were no drinking water facilities, as the canals had not yet been completed. Whenever rains failed, water was arranged by the authorities and supplied through tankers loaded on bullock carts. In the years when rainfall was reasonably fair, rain-fed cultivati n was done even though the yield of crops was quite low. Bajra, jowar, gram, and barley were the main crops. No wheat could be cultivated then.

In rainless years, the migration of people and cattle to greener areas under the old canal system became inevitable. Though wells were constructed, the groundwater was brackish and available at depths of 50 metres or more. However, in a few villages like Ram Nagaria and Bhamson in Sirsa, fresh water was found. Electricity came to a few villages in 1961. There were no schools and so, education facilities for children were non-existent in the first few years. Before they built their own houses, the displaced persons were accommodated in temporary shelters. In addition, no proper cattle shelters were available.

PRESENT CONDITIONS

Those allottees who braved the initial hardship are now quite well-off. Most of them have *pucca* houses. As with other allottees (many of them being displaced persons from Pakistan who also had to start life afresh), they are also harvesting good rabi and kharif crops. They have reaped the benefits of the 'Green Revolution' ushered in through HYV seeds, chemical fertilizers and irrigation. Now they have most of the essential facilities like a primary school and health centre in the village, pucca roads, tractors and other mechanical devices for ploughing, harvesting, etc. They employ labour from outside the state to help them in farming. While the older generation was nostalgic about their previous villages, the younger generations have adjusted to the local environment and are fairly contented. However, problems relating to lack of sufficient water supplies in the canals, prices of agricultural products, concern about water tables, and so on, were voiced. These people expressed a strong desire to improve their economic lot. They have, however, continued to maintain their old social customs, language, and attire.

Beneficiaries of Irrigation

Assured supplies from the Bhakra dam storage have helped not only to stabilize and improve the existing run-of-the-river system, but have also opened up new areas to irrigated agriculture in Punjab. Thus, all the areas lying on the eastern side of the Sutlej up to the Ghaggar were covered with irrigated agriculture. Similarly, the *doab* between the Beas and Sutlej, which had earlier been practising groundwater-based agriculture, now has the benefit of assured water from the Bhakra system. In the case of Haryana and Rajasthan, the Bhakra canal network opened up virgin but semi-arid and arid desert areas to the benefits of irrigation and water supply.

The questionnaire drawn up for the farmers in the command area had sought information generally with respect to the following aspects:

IRRIGATION

- Irrigation coverage in Punjab

    (i) Existing canal system—Sirhind canal system in Punjab
    (ii) New canal systems—Bist-Doab canal system in Punjab
    (iii) Other new irrigation areas in Punjab

- Irrigation coverage in Haryana
- Irrigation coverage in Rajasthan
- Availability of power for pumping
- Adequacy of supply and timely intimation of shortages
- Effects of canal supplies on groundwater

AGRICULTURE

- Crops grown and cropped area
- Adoption of modern agricultural technologies
- Assistance from research institutions/universities
- Source of extra labour
- Existence of water users' associations
- Dairy development

SOCIAL AND ECONOMIC CONDITIONS

- Drinking water supply and improved health status
- Education
- Employment opportunities
- Infrastructure developments
- Development of industries
- Improvements in standard of living

IRRIGATION

Irrigation Coverage in Punjab

(i) *Existing Canal System—Sirhind Canal System in Punjab.*

The Sirhind canal system that emerged from the Ropar headworks and fed by 'run-of-the-river' supplies of the Sutlej was the only existing canal system in Punjab before the completion of the Bhakra dam.

Owing to additional water being made available through the Bhakra storage and the enlargement of the canal capacity of the old system by over

99 cumecs, new areas were added to the existing system. This resulted in the remodelling and redesigning of existing outlets as well. Originally, these lands were being irrigated from open wells by the Persian wheel lifting system.

Areas along the Sutlej river were frequently inundated in the past during flood, and these reclaimed areas were brought under irrigation through the Sidhwan branch and the Abohar branch. In these areas, both canal and tube-well irrigation are now being practised. Tubewells along the canals have sweet water, while others are slightly saline. The farmers are quite aware of the water quality because water-testing facilities are readily available. They often resort to blending or alternate irrigation by canal and undergroundwater. Areas in the upper reaches of the Bhatinda branch and the Kotla branch generally use surface waters, while sub-surface waters are used during times of scarcity. Paddy cultivation has necessitated more extensive use of groundwater to supplement canal supplies. In the lower reaches, particularly in the Bhatinda branch, farmers have to rely only on canal water, as the undergroundwater is brackish and not suitable for irrigation. They complain about inadequate water supplies to meet their present needs and consider this a major deterrent for their economic betterment to levels of those such as the Patiala and Ludhiana districts.

In the water-deficient areas of the Bhatinda district and elsewhere in Punjab, cotton cultivation is popular.

(ii) *New Canal Systems—Bist-Doab Canal System in Punjab*
The area situated between the Beas and the Sutlej, and popularly known as Bist-Doab, was earlier being cultivated by drawing water from open wells and was considered fertile. However, due to the extensive overdraft of groundwater, the area started experiencing lowering of the water-table. This necessitated the decision to provide canal irrigation for the region.

This step has since enabled the cultivators to harvest bumper crops. Both sources of water are being exploited now, and, in line with the trend of changing cropping patterns noticed in other parts of the state, these farmers have also gradually switched over to the cultivation of paddy as the principal kharif crop and wheat as the rabi crop, with highly favourable economic returns.

The farmers pointed out that in the recent decade or so, the maintenance of the canal system had been neglected. Siltation in the distribution system and paucity of canal waters at the tail of the canals were the result, they alleged. They had to, therefore, resort to extensive tube-well irrigation. However, this led to lowering of the water-table, requiring deeper boreholes and greater maintenance costs. Presently, per capita landholdings have come down, making tubewell irrigation unviable.

The farmers feel that there is need for proper maintenance of the distribution system. The cultivators showed greater preference for canal waters as, in their perception, it was of better quality and more economical. They also asserted their right to receive additional canal water.

(iii) *Other New Irrigation Areas in Punjab*

Areas along the Ghaggar river which were being irrigated mostly by inundation irrigation during floods were provided irrigation through a network fed by the Bhakra main branch system. Cultivators pointed out that before 1960, while periodically severe floods caused an almost total loss of kharif crops and cattle, subsequent rabi crops used to be good and would help them to partially recover their losses. Some places had Persian wheels where water was available at 5 m depth or so. However, they were very happy when assured supplies were received from the Bhakra system. The area has since been protected from flood. These areas have also started using groundwater to supplement their increased requirement as a result of switching over to paddy cultivation. Some cultivators expressed the hope that the proposed Markanda dam would help them in the future to secure more water for irrigation.

Irrigation Coverage in Haryana

Except for some irrigation along the Ghaggar river, all other areas in Haryana now covered by the Bhakra canals were originally barren, sparsely populated semi-arid areas. Rain-fed irrigation for cultivating bajra and jowar was sometimes resorted to but the yields were very poor. Efforts were directed mainly towards conserving water for drinking purposes. The groundwater was generally brackish and unfit for irrigation. Large areas were desert land with sand dunes similar to the neighbouring areas of Rajasthan. When the rains failed, there was migration of both people and cattle to the greener areas of neighbouring states.

Bhakra irrigation has now helped the cultivators to lead a settled life. In the initial operative years of the canal system, there were problems of sand casting of canals, frequent breaches and so on. The irrigation system stabilized after some years. By this time, the farmers had gained good experience in irrigated farming. The general complaint voiced was that Bhakra supplies were below their current requirements and sufficed for about 50 per cent area because the original water allowance was low.

Except for a few pockets, underground water in most areas of Haryana covered by the Bhakra canals is brackish and not good for irrigation or drinking. In all these areas, the introduction of good quality canal waters has helped raise the levels of the groundwater table. Farmers have also taken up cultivation of paddy in a big way. Almost 80 per cent of their crop in kharif is paddy. It is only the conjunctive use of canal and groundwater that enables them to grow this paddy crop. Cotton, bajra, pulses, mustard, barley, and gram are some other crops popularly grown in rabi.

Irrigation Coverage in Rajasthan

For farmers of the Hanumangarh district of Rajasthan, which lies at the tail-end of the canal system, the Bhakra canal water is the only source of water available for *any* use. The canal has become a 'lifegiver' to them and they refer to it as the *Maru Ganga* (Ganga of the desert).

The area was completely barren before 1960 and the population was very sparse. There was only cultivation of coarse grains like bajra and gram during the years when it rained. A little shower was, however, good enough for grass to grow for cattle grazing. Hence, cattle-rearing was the main occupation before 1960. The cattle had to be, however, moved to greener canal areas in the event of failure of rains, which was very frequent. Almost four out of five years were practically dry, leading to such migration. Even drinking water had to be brought by trains. The heavily brackish groundwater was unfit for drinking as well as cultivation.

The situation improved dramatically after the supply of water from the Bhakra canals. The cultivators stressed that they owe their very survival to these waters.

Availability of Power for Pumping

All villages in the command area in the three states receive power both for domestic and agricultural purposes. Groundwater is invariably used in addition to canal waters. Farmers generally use electricity to operate their pump. A constraint for these farmers is the limited number of hours (8 to 12) for which the supply is available in rural areas. Therefore, a substantial number also used diesel for running the pumps. Small and marginal farmers found themselves unable to bear the cost of taking electric connections to their tubewell sites.

In the lower reaches of the Bhatinda branch and a few villages of Sirsa and Hanumangarh, the groundwater was brackish and, in such places, the question of pumping groundwater did not arise.

Adequacy of Supply and Timely Intimation of Shortages

Though not fully satisfied with the overall quantum of waters made available, the farmers were, nevertheless, happy that their allotted share of water was assured. There is an efficient system of *warabandi,* and shortages are intimated well in advance and equitably shared amongst the beneficiaries of any outlet.

Effects of Canal Supplies on Groundwater

In most areas, the depth of the water-table before 1960, when canal supplies commenced, ranged between 20 to 30 metres, and the groundwater was brackish as well. Perennial irrigation over the years changed the position and raised the water-level. Drainage congestion caused by road and rail network, as well as the poor natural slope coupled with seepage from unlined canals, led to the rise of the water-table in many areas. Many low-lying areas became waterlogged within a few years of the introduction of canal irrigation. This menace was immediately recognized and a programme of drainage improvement, including lining of the distribution system in the vulnerable areas and encouragement for the construction of tube-wells, was rigorously implemented. The concept of vertical drainage through tube-wells supplemented by horizontal drainage soon helped achieve remarkable results, and lowered the water-table. Now the farmers feel that there is *no* problem of waterlogging or salinity. None of the areas visited had any indication of waterlogging or salinity problems, nor were there any complaints in this regard.

AGRICULTURE

Crops Grown and Cropped Area

In the head reaches of the canal system where the canal water supply is considered by farmers to be satisfactory and the availability of groundwater is good, wheat is the principal rabi crop and rice that of kharif. Invariably, all the available land is put under the plough for both rabi and kharif, along with the use of chemical fertilizers and HYV seeds. However, farmers in the tail reaches of most canals voiced complaints of insufficient canal water for irrigation.

In areas where the groundwater is saline, rain-fed gram is popular. Mustard is being grown as a rabi crop mostly in the Sirsa and Hanumangarh districts. American cotton, locally known as *Narng,* has been less popular over recent years. The farmers attributed their hesitation to its being frequently affected by pests which destroy the buds very quickly. However, in 2003, which saw good monsoon rainfall, there were very satisfactory levels of yields with respect to all crops. Cotton was also flourishing in areas of the Bhatinda district served by the Bhatinda distributary of the Sirhind canal. Another reason attributed currently by farmers for the loss of popularity of cotton is the non-remunerative price.

The average yields indicated by farmers were—wheat: 16 to 20 quintals/ acre; cotton: 6 to 8 quintals/acre; paddy: 20 to 25 quintals/acre.

The cropping pattern had undergone a marked change over the years, with paddy and wheat becoming the main kharif and rabi crops, respectively. Jowar, bajra, and maize, which were dominating crops in the pre-canal era, are now being grown sparingly and merely for local consumption.

Adoption of Modern Agricultural Technologies

At the time of Independence, the districts of Hisar and Sirsa, now in Haryana, had practically no irrigation facilities. The Bhakra canals not only helped bring the new areas under irrigation, but also improved the network by helping to recharge the groundwater. The BNP provided electricity to lift water economically through the rural electrification network. Bhakra irrigation enabled the states to implement a policy for promoting the use of HYV seeds, chemical fertilizers, and mechanization of agricultural operations. The area turned into a 'bread basket' and brought about the 'Green Revolution'. Now, the farmers served by the BNP have acquired expertise and show keenness to adopt new technologies in agriculture with the assistance of scientists in research stations and universities.

Progress in the introduction of sprinkler/drip irrigation techniques is very slow. This is attributed by farmers to the lack of incentives and the high cost of installation, as well as the slackening level of agricultural extension services in this respect. These new technologies are being tried with success only for horticultural development, where farmers are satisfied with minimal requirements of water for better output of fruit crops. The promotion of these water conservation technologies will help achieve higher levels of productivity for farmers. Setting up demonstration farms in the area may also help achieve greater acceptability by them.

Farmers generally emphasized the fact that the introduction of HYV seeds, pesticides, fertilizers, etc., for high production would not have been possible without Bhakra water. Punjab and Haryana would have remained semi-arid tracts, except for some modest and uncertain inundation irrigation. They pointed out that without the widespread development of surface waters, the recharge of the groundwater and, thus, the promotion of tube-well irrigation to supplement surface water availability, would not have taken place.

Assistance from Research Institutions/Universities

Many farmers expressed the view that agricultural universities/state departments did not extend much help directly, except for holding *kisan melas* periodically in various areas. There were no demonstration farms which could have shown the effectiveness of advanced techniques of agriculture and irrigation management.

Source of Extra Labour

Most of the farming work in Punjab is done by using labour brought from Bihar, eastern Uttar Pradesh, and Orissa, particularly for harvesting. Mechanical operations like running the tractors, harvesters and so on, were mostly done by the farmers themselves. Labour from outside is being used in Sirsa/Hisar in Haryana, and Hanumangarh in Rajasthan to supplement local hands, but not to the extent to that it is being done in Punjab.

Existence of Water Users' Associations

The warabandi system exists in all the areas served by the canal distribution system. Under this system, the water flowing from an outlet is given to one farmer at a time by rotation for a predetermined fixed period, which is proportional to the size of his holding. The sequence and duration of turns are specified in a roster prepared by the canal officer in consultation with the stakeholders. This system has been in operation for many decades in all irrigation systems developed in north-west India. The concept of Water Users' Associations (WUAs) which assume responsibility for equitable distribution of water, maintenance of canals, collection of water revenue, sanction of irrigation water for other uses and so on, is being tried by various states of India. This has not yet become popular in the BNP area. There were no WUAs in any of the villages of the BNP that were visited.

Dairy Development

In the arid and semi-arid regions of Rajasthan, cattle-rearing is popular. The easy availability of fodder after the BNP has encouraged farmers in all the three states to keep larger herds of cattle like cows and buffaloes for milk production, both for domestic and commercial purposes. Many private and public dairies which cater to urban requirements have sprung up in the BNP area.

SOCIAL AND ECONOMIC CONDITIONS

Drinking Water Supply and Improved Health Status

Open wells served as the sources of drinking water before 1960. At present, all villages have a water supply system with piped water connection in most houses. When a village is located near a canal, water is drawn from it directly. If the groundwater is sweet and easily available, tube-wells also become the source. Village ponds used to cater to cattle needs earlier but the cattle are now served by the regular water supply system. Urban areas and towns located in the command areas have a municipal water supply with treatment plants. Canal water has helped meet the drinking water requirements of most towns and villages in the BNP command.

Sanitation has also greatly improved as compared to the pre-1960 era. Better quality of water and general improvements in sanitation have helped control diseases like malaria, filariasis, and so on. The farmers stated that water-borne diseases were very common before 1960. Besides, there was hardly any health-care system then. Primary health centres now exist in all villages.

Education

While some villages in Punjab had primary schools, there were hardly any schools before 1960 in the rural areas of Haryana and Rajasthan. There is now a marked change in the attitude of farmers towards the education of their children. Most of their children are sent to school, as primary education facilities are now available in all villages. High schools are generally available within 5 to 10 km, while colleges are in towns and cities. Farmers now consider that with better education, their children will acquire better skills in farming. They will be able to deal with government departments and the administration with confidence. It will also enable them to secure better returns on their produce and, thus, achieve betterment of their economic conditions.

Another feature noted was that girls were also attending schools in larger numbers in all villages. However, this trend is more noticeable in Punjab and Haryana than in Rajasthan.

Employment Opportunities

The development of irrigation has resulted in the increase of employment opportunities. Local labour is insufficient to support all the agricultural operations resulting in the need for hiring from other states. New townships, mandis and transport systems (public and private) have sprung up. All these have led to an increase in employment opportunities, both for educated and uneducated youths. The all-round development triggered by irrigated-agriculture has been a boon for the unemployed and underemployed, as also the landless categories of people.

Infrastructural Developments

Road development was very poor before 1960. Only katcha roads and paths existed. There has been remarkable development in this respect in the past few years. All villages now have pucca bitumen roads, which helps in linking the villages to towns and mandis by regular bus services. Bullockcart-based transportation has now been replaced by the tractor-trolleys, which ferry agriculture produce to the mandis.

Development of Industries

Cotton ginning and spinning mills have come up at many places like Hanumangarh and Bhatinda. Their number is, however, not large. Similarly, there is enough untapped potential for food and fruit-processing industries. Farmers were quite conscious of this and urged that the authorities should promote such agro-based industries.

Improvements in Standard of Living

There has been much improvement in the standard of living of the farmers and other rural communities. With increased income levels due to bumper crops, the farmers were able to construct pucca houses. Presently, there are hardly 5 to 10 per cent katcha huts in the Punjab villages. The level of prosperity seen in the Bist-Doab system and the upper reaches of the Sirhind canal system, is higher than that in other areas of Punjab. With increased education, one or two members of each family in many villages in Jalandhar and Patiala have opted for jobs in civil and defence establishments and are doing very well. Many have even migrated abroad. With additional funds remitted by them back home, farmers in villages in Punjab have constructed very good houses, much better than those seen even in affluent urban areas. They have all the modern amenities like televisions, refrigerators, piped water supply, and water-borne sanitation systems. With pucca roads available, buses, taxis and other means of transport have bridged the urban and rural divide. All villages are electrified. Telephone services have penetrated even to remote areas. Cellphones have also become quite popular.

There has been a marked change in the food habits of the rural population after the advent of irrigated agriculture from the BNP. Farmers like those in the Hanumangarh area, as well as in adjoining districts of Haryana, whose staple diet used to be jowar and bajra, have now switched over to wheat. Flavoured basmati rice is a popular addition to their menu now. With plenty of locally available vegetables and dairy products, there are welcome improvements in their ways of living and eating.

14

SOME IMPORTANT LESSONS LEARNT

Need for Post-Project Performance Analyses

Constant efforts are needed to assess the manner in which water resources development projects actually perform during their useful existence and operation. For this purpose, we must periodically assess completed projects and judge the impact of each of these schemes with regard to various aspects. These need to be compared with the targets proposed and the impacts anticipated at the stage of their approval for execution.

India has completed around 4000 water resources development projects involving the construction of large dams. However, there are hardly any well-researched reports assessing their actual post-project performance. What usually gets projected as the impact of dams are either the anticipations of the project planners or the apprehensions about the possible adverse effects voiced by critics. One major reason for this deficiency may be the lack of data for such studies. Whenever studies on the performance of projects are initiated, they are confronted by the non-availability of requisite data relating to both 'pre-dam' and 'post-dam' conditions in the project area.

The expert committee, which recently reviewed the implementation of the recommendations by the National Flood Commission, noted the paucity of performance analysis studies of past schemes. It asked for the analyses of performance of schemes including the impact on the socio-economic development of the area benefitted. It also noticed a tendency to dilute the importance given to certain aspects by giving preference to others through modifications in regulation arrangements over the years.[1]

Fortunately, much of the basic data in respect of the BNP region have been collected and are available in one form or the other with different

---

[1] Government of India, Ministry of Water Resources, Report of Experts Committee on Review of Implementation of Recommendation of Rashtriya Barh Ayog, March 2003, para 4.5.

agencies. Thus, the present study has been able to make real headway. The assessment of performance has clarified that the project has fulfilled all the stated objectives in a sustained manner. In addition, it has provided immense additional benefits to the region through indirect and secondary impacts. It was concluded that the BNP is a successful story of sustained humane development with overwhelmingly beneficial impacts for the region and the nation. What we shall now attempt is to indicate some important lessons learnt from this study on issues that would have wider applicability in most projects. Similar studies need to be done in respect of the many other projects that have been in service for decades.

Non-Transparent Governance

Almost every water resources development project involving the construction of large dams undertaken in India has been planned, built, and operated by the state governments. These government departments and the agencies and undertakings under them are also subject to the same ills commonly observed in public governance—particularly their non-transparency. As a result the public and the NGOs are invariably unable to obtain all the relevant data and details about the specific manner in which such projects were planned, approved, constructed, and operated. Basic data and details gathered are often locked up within the labyrinth of the government and not made public. The 'right to information', accepted in principle, is of no use in many cases. In cases where certain sensitivities are involved—such as those impinging on interstate issues or international drainage basins—much data is of a classified nature, unavailable to researchers or to the public. Often, this leads to vested interests putting out 'suitable' data or details to serve their purpose, or to unverified or even patently wrong information becoming the 'knowledge base', which then is used for drawing incorrect conclusions.

The only way in which this situation can be corrected is for the government to either make all data on past projects easily available to the public, or at least sponsor periodical performance reports through credible institutions, inside or outside the governmental system.

Planning for Power

It was mentioned earlier that the Planning Commission and the government were reluctant to approve all the units of hydropower stations under the

Bhakra dam, apparently fearing a glut of power and locked-up investment. The Planning Commission was aware that the per capita electricity generation was hardly 18 kWh in 1951, and that it rose to 28 kWh by 1956, and stood at 45 kWh in 1961.[2] Was this hesitation justified, particularly when the Bhakra project was for relatively cleaner hydropower development in a power-hungry region?

The World Development Indicators 2003, published by the World Bank, highlights the basic fact that 'economic growth and greater energy use have a direct and positive link, evident in the rapid growth of commercial energy in low and middle income countries'.[3] Even high-income countries use more than five times as much as the developing countries on a per capita basis, and with only 15 per cent of the world's population they use more than half of its commercial energy. The difference in per capita energy use can be striking: the USA, for instance, uses sixteen times as much energy as India.

It has been recognized that access to electricity and its use are important in raising the standard of living of the people. Where does India stand now among the nations of the world in this regard? The Human Development Report 2003, published on behalf of the United Nations Development Programme (UNDP), indicates the electricity consumption per capita at the turn of the last millenium in the year 2000, with respect to 175 countries. The world average per capita consumption was 2176 kWh. India uses 355 units (kWh) per capita per annum, which is a mere one-sixth of the average world use. The use by high-income countries averages 8651 units. The Indian use is less than half of the average for developing countries. Many high-human development, high-income countries like Norway, Sweden, the USA, and Canada, now use well over 10,000 units of electricity. It should be noted that three of these were using more than 10,000 units, per capita, even 23 years ago, in 1980. The USA, for instance, was using nearly 9000 units in 1980.

A comparative picture of the relative per capita consumption in 1980 and 2000 in some countries is provided in Table 14.1.

The per capita consumption of electricity, again, varies among the states of India. Punjab was always at the top of the list, with Haryana not far behind. The position over the years is shown in Table 14.2. However, even Punjab, with the highest per capita consumption among all the states of India, consumes less than half of the world average.

[2] The Planning Commission, *Third Five Year Plan*, 1961, para 49, p. 399.
[3] World Development Indicators, 2003, World Bank, Washington D.C., USA.

Table 14.1: Growing electricity consumption in some countries

Country	Per capita consumption kWh		Population 2001 millions	GDP US$ 2001 per capita
	1980	2000		
Developed countries				
Norway	18,289	24,422	4.5	36,815
Sweden	10,216	14,471	8.9	23,591
Australia	5393	9006	19.4	19,019
USA	8914	12,331	288.0	35,277
Canada	12,329	15,620	31.0	22,343
Japan	4395	7628	127.3	32,601
Switzerland	5579	7294	7.2	34,171
Denmark	4222	6079	5.3	30,144
UK	4160	5601	58.9	24,219
France	3881	6539	59.6	22,129
Germany	5005	5963	82.3	22,422
Developing countries/LDC				
India	130	355	1033.4	462
China	253	827	1285.2	911
Pakistan	125	352	146.3	415
Nepal	11	56	24.1	236
Bangladesh	16	96	140.9	350
Egypt	380	976	69.1	1511
South Africa	3213	3745	44.4	2620
Malaysia	631	2628	23.5	3699
Brazil	975	1878	174.0	2915
Turkey	439	1468	69.3	2230
Thailand	279	1448	61.6	1874

Source: Human Development Report 2003, UNDP, Oxford University Press.

Table 14.2: Per Capita Consumption of Electricity (kWh)

State/Union Territory	1970–1	1980–1	1989–90	1999–2000
Chandigarh	280.2	309.0	686.2	823.8
Punjab	156.2	303.6	620.5	921.1
Delhi	250.6	403.8	673.6	653.2
Haryana	88.8	209.5	367.4	530.8
Rajasthan	36.8	99.4	191.6	334.5
India	79.8	120.5	236	354.7

Source: Tenth Five Year Plan (2002–07), Planning Commission, Govt. of India, vol. III, p. 87.

In retrospect, the Government of India has been slow in correctly assessing the likely demand (nationally or region-wise), or in realizing the cost of delay in sanctioning schemes to meet the demands. Many well-meaning economists and environmentalists, too, have made it more difficult. Perhaps, raising the requisite funds for investment was also a problem due to policies adopted that were not conducive to it. Surprisingly, many of the countries in the Himalayan region, which have significant hydropower potential, have not fared any better than India in the timely conversion of that potential into reality. They do not seem to have learnt from India's experience in this regard, or from the opportunity costs of delays in such development. For the people of these countries, candlelight dinners have become part of the humdrum everyday existence, instead of being redolent with romance! The greater irony is that some of the great lobbyists and the support for stoppage of such development in the least developed and developing nations come from countries that have nearly completed similar tasks and are enjoying the benefits thereof. The government and the people of the USA were reminded by the blackout of August 2003, of similar occurrences in 1965 and 1977. For people in developing nations like India, it is the grim replay of their everyday grind. It would help steel the will of these nations if they refused to tolerate meekly their present saga of 'living with enfeebling disadvantages' like water and power shortage or disease, poverty and misery.

Integrated Drought and Flood Management

One of the surprises in the planning of the BNP was that although it aimed at drought mitigation of vast areas in Punjab as a primary objective, in addition to hydropower generation, flood mitigation was relegated as an incidental benefit. While it is true that Punjab was seriously affected by famine in many drought years, it was also the scene of devastating flood in the past. These were found to be no less injurious to the economy of Punjab than droughts. However, it is gratifying to note that the importance of incorporating flood management too as an important objective has since been recognized and reflected upon, in some measure, in the present operation of the reservoir. Integrated flood management is now viewed as the best way of optimizing the various objectives and potential of a land and water resources development project. It is necessary for the states involved in the BBMB to re-examine the present situation in the light of fifty years of post-dam experience, and agree on the re-ordered priorities and optimal manner of operation of the integrated system.

This will also be in tune with the recommendation made by the Working Group for flood control for the Tenth Plan.[4]

The Working Group stated as follows:

Multi-purpose reservoirs often involve a compromise of varying interests. The interests of various components like irrigation, power generation and flood control are often at variance with one another and therefore a compromise is made in order to realise the optimal benefits for the project as a whole. A reservoir is more effective for flood control if a designated space is reserved and a reservoir regulation arrangement is laid down....

...The working group would suggest that during the Tenth Plan period an expert group may be set up to review the major dams in India and suggest if the reservoir regulation can be modified for better flood management in the future.

(Paragraph 5.3.5)

Drinking Water Supply

If water is essential to sustain agricultural growth and productivity, it is even more vital for life and healthy living. Over half of morbidity cases arise from drinking impure water. The toll in mortality, debility, and productivity is grievous. Most irrigation and multi-purpose projects of the past had not specifically mentioned that providing drinking water to the region served by them was a primary objective. Nevertheless, all such projects have actually served as the source of raw water supply for the thousands of towns and villages in the command area. Whether highlighted or not, they were invariably pressed into rendering this service as well. Bhakra-Nangal is no different.

Planners of Indian water resources projects have always considered drinking water supply a basic need and included the cost but did not give it a monetary value among the benefits of the projects. Yogender K. Alagh, a former minister of the Government of India who took an active part in the planning of the Sardar Sarovar project, has highlighted this aspect.[5] He notes that the opportunity cost of provision for reliable sources of drinking water is very high but there is a dilemma in assigning a monetary value for 'life' (see box—'Swim out of ghost waters').

[4] Government of India, Ministry of Water Resources, Report of the Working Group on Flood Control Programme for the Tenth Five Year Plan (2002–7), August 2001.

[5] Alagh, Y.K., the *Indian Express*, 30 December 2003.

Excerpt from the

Indian Express

30 December 2003

SWIM OUT OF GHOST WATERS

GRAIN OF TRUTH

Yogender K. Alagh

... The first argument given is that the original project did not have a drinking water component. But that is wrong. It is true that the original published plan did not account for the benefits of the drinking water component. As I had explained then, this emerged from the fact that the SSP planners were good students of Indian planning. In those days drinking water was considered as a basic need and so you only counted the cost and did not give it a monetary benefit value since how do you value 'life'?

The need for drinking water for people and cattle in many areas of Punjab and Rajasthan was well known. An interesting scene witnessed by the 'Working Party' involving India, Pakistan, and the World Bank in December 1952, during the Indus Treaty negotiations, is revealing. N.D. Gulhati, who was present there as India's representative, records:

There was, however, one interesting unrehearsed demonstration of water scarcity in one location which could not but have had a profound effect on the visitors. It was the last day of December 1952. The Working Party was scheduled to go to Ganganagar along the Otu canal. On reaching the Otu headworks, we found that there had been some unexpected rain and the earth road along the canal was too wet to be motorable. Accordingly, it was decided to catch a local train for Ganganagar from a nearby railway station. As we arrived at this station, there was a long queue of village folk, mostly girls, who had come from miles around to get household water from the only source—a water tank at the station brought by rail from some distant place. Until the train arrived, about three-quarters of an hour later, village girls in their colorful dresses kept on coming to the water wagon to get drinking water for their families—an interesting but heart touching spectacle such as the visitors had never seen before.[6]

[6] N.D. Gulhati, *Indus Water Treaty*, Allied Publishers, Bombay/Delhi, 1973, p. 112.

The BNP project report did not indicate water supply as a primary objective. However, when people of the project area were interviewed for the present study, they brought out forcefully the fact that the BNP served as a very important source of their drinking water supply. This benefit was no less important to them than its irrigation facilities.

In many parts of India, especially in the hilly and semi-arid regions, women, from their earliest childhood, are condemned to hard labour and drudgery, fetching water for the family. Hence, provision of potable water for their homes is a major step, too, in rendering gender justice to them.

All future water resources development projects should give due recognition to this aspect right from the early stages. Water supply to people in the benefitted area for domestic use should invariably be made a primary objective.

Importance of Maintenance

Water resources development schemes have a long life. In India, many of them have been performing satisfactorily for over a hundred years. This brings into focus the need for satisfactory maintenance of the various facilities created at sizeable initial investments.

The Bhakra dam is now four decades old. The Nangal barrage and canal system have functioned for five decades. Complaints about inadequate maintenance of created assets have, however, been voiced in recent years. It is a fact that states have a tendency to give great importance to the creation of new infrastructure works without paying commensurate attention to the adequate maintenance of created assets. Inadequate provision of funds for this purpose and the continuous neglect—sooner or later—leads to serious situations.

There has certainly been less than required attention paid to the maintenance of the extensive canal system created by the BNP. So, too, is the case with most multi-purpose river valley projects of India. Each project needs to be closely examined on this aspect and remedied to ensure trouble-free performance.

Trauma of Displacement

Construction of infrastructure projects like large dams provides immense benefits but there are some adverse social impacts, too. The submergence of homelands and habitation causes involuntary displacement to some people who, thus, become 'oustees'. Additionally, there can be occupational and cultural displacement. The most significant social cost of large dams is borne by these people. The financial cost of relocating and resettling them becomes a part of the project cost. However, people's attachment to their ancestral homes and their communities, as also their sentiments, cannot be really assessed in monetary terms. This fact underlines the need to consider the displacement issue with the utmost care and sympathy.

An oft-repeated query is why people, particularly tribals, should be involuntarily displaced at all, especially if the costs are to be borne by them and the benefits by others. The fact is that displacement is a continuous process, and not by any means unique to dam projects. If development sometimes displaces local populations, including tribals, so does lack of development. Poverty, aggravated by increasing population pressure, has forced hapless communities to erode their natural capital resource base for their very survival, compelling out-migration in due course of time. The 'distress migration' from the countryside is increasingly filling shanty towns. This 'rural-urban push' does not respect tribes or tradition. Lack of development is displacing a considerable number of people annually.

The effort of the government has generally been to prepare a Relief and Rehabilitation (R&R) package seeking to ensure that the oustees do not become worse off because of migration. In the early reservoir projects undertaken by independent India such as the BNP, this aspect did not cause any real problem. In any case, there are no past records of any such serious problems. In many cases, there is not even a reliable and comprehensive record of what happened to the displaced people. However, in the case of Bhakra, details of all those who were given compensatory land and housing plots in the Hisar district of the irrigation command are available, and the deputy commissioner of that district is the designated authority to deal with these issues even today. Yet, even in the BNP, no details are available about the urban displaced of the Bilaspur town, or of the rural oustees who chose to receive their compensation in cash and resettled on their own.

In many recent projects (like the Sardar Sarovar or Tehri), more comprehensive and fairly liberal R&R packages have been provided. The basic principle followed in the recent R&R packages is that the economic condition of the displaced people should be no less than what was enjoyed by them earlier. Nevertheless, complaints have persisted that even in those progressive packages, many needs are inadequately or inappropriately provided for. Apart from the inadequacy of the packages, complaints are made about their implementation.

National Policy on Resettlement and Rehabilitation

A National Policy on R&R was drafted in 1998 and widely discussed with a view to its finalization and adoption by consensus. The government finally notified the policy in February 2004. However, it has been criticized for diluting many desirable features which existed in the earlier drafts. Therefore, the much sought after consensus seems to be still elusive.

The National Policy on Resettlement and Rehabilitation for 'Project Affected Families' 2003 (NPRR-2003), which became effective from 17 February 2004, is applicable to all projects displacing 500 families *en masse* in plain areas, or 250 such families in hilly areas. States where the R&R packages are higher than those mentioned as minimum in the National Policy are free to adopt their own packages.

The broad objectives of the policy are:

- to minimize displacement
- to plan the R&R of project-affected families (PAF)
- to provide better standard to the PAF; and
- to facilitate a harmonious relationship between the project authorities and the PAF through mutual cooperation

The steps prescribed include the:

(i) designation of an administrator and commissioner for the R&R;

(ii) identification of families likely to be affected, in a time-bound manner;

(iii) preparation of a list of lands that may be availabe for the R&R of the PAF and deciding the resettlement area;

(iv) acquisition of land under the Land Acquisition Act (1894);

(v) drawing up and notification, after due publicity, of the scheme for the R&R, in consultation with the representatives of the PAF;

(vi) setting up of dispute redressal mechanisms; and

(vii) monitoring of the progress of implementation.

The full details are given in the Resolution dated 17 February 2004 of the Ministry of Rural Development (Department of Land Resources).

Let There be no Losers

The ideal to strive for is a situation where there will be no adversely affected people or 'losers'. Efforts should be made to make the PAF beneficiaries of the scheme. Many suggestions in this regard have been made at different times. A surcharge or cess could be levied on the energy produced. A betterment fee could be collected from the benefitted farmers and communities and a fund set up for the betterment of the 'project-affected'. This fund could be spent for the catchment area improvement, provision of education/training/skill formation, for the development of fisheries and markets, for setting up agro-processing industries, and so on in the upstream catchment. Goals for achievement should be set up right at the project stage and their implementation should be monitored by the government. Perhaps a catchment area authority could be set up to take charge of this. The other steps that could be considered include the provision of opportunities for the oustees to purchase irrigated land on a priority basis, provision of electricity, water supply, and so on from the project, as also appropriate training support for the self-employed.

Public Relations and People's Involvement

People are central to economic and social development. India, functioning under the democratic system of governance, must give due importance to the views of people and continuously interact with them with respect to various development projects. Major water resources development projects of India involve the lives of millions of people in their day-to-day routine. Thus, the need for people's involvement at all stages is crucial. In fact, the business of governance is too important to be left to governments alone. Enlightened citizens can help penetrate the fog of narrow partisanship and enable public opinion to recognize where their real interests lie.

The state and central governments and their agencies can do a lot in maintaining an effective interface with the public. The public relations aspect of governance, too, needs polishing up drastically. Governments are often too slow to react to people's perceptions, unless they acquire the form of serious agitations. There is need for introspection in this regard.

Impact of Large Dams: Summing Up

The Nangal part of the BNP has been in the service of the nation for over five decades. The Bhakra dam was completed over four decades ago. The assessment of performance has revealed that the project has fulfilled all the stated objectives in a sustained manner. In addition, it has provided immense additional benefits to the region and the nation through indirect and secondary consequences. The perceptions of the affected people, too, were ascertained through random interviews with them. The broad conclusion emerging from these is also that the project brought immense prosperity to them.

In the case of the BNP, the gain to the poorest group was much higher than that for the average, and even higher than for the land-owning farmers. Such a result signifies an important implicit message, that the dam acts as a powerful vehicle for poverty alleviation.

The general lesson learnt from the study of the BNP is that large dams, apart from their direct impact, have a significant indirect and induced impact as well, which must be accounted for and taken into consideration in any project evaluation.

Bhakra-Nangal is a success story of sustained humane development with overwhelmingly beneficial consequences—clearly evidence of the importance of large dams. All in all, it stands tall as concrete testimony to the far-sighted vision and wisdom of the pioneer dam builders of independent India.

Annexures

ANNEXURE 1

List of participants who attended the workshop on 12 January 2004 at the Centre for Policy Research, New Delhi

CENTRE FOR POLICY RESEARCH

1. Charan Wadhva
   President and Chief Executive

2. R. Rangachari
   Research Professor and Project Coordinator

3. M.L. Baweja
   Former Chief Engineer (CWC)

4. B.G. Verghese
   Research Professor

5. Ramaswamy R.Iyer
   Former Secretary (Water Resources)

6. Nawal K. Paswan
   Senior Research Associate

NIPPON FOUNDATION, TOKYO

7. Masonari Tamazawa

THIRD WORLD CENTRE FOR WATER MANAGEMENT

8. Asit Biswas
   President

BHAKRA BEAS MANAGEMENT BOARD

9. Rakesh Nath
   Chairman

10. Balbir Singh
    Member (I)

11. Anil Arora
    Secretary

12. D.S. Notra
    Chief Engineer, Bhakra Dam

13. H.C. Chawla
    Director (C)

14. Rajan Bansal
    Executive Engineer (to Member I)

15. K.V.S. Thakur
    Director (Regulation)

MINISTRY OF WATER RESOURCES

16. R. Jeyaseelan
    Chairman
    Central Water Commission

17. B.D. Pateria
    Chief Engineer (Project Monitoring Organization)
    Central Water Commission

PLANNING COMMISSION

18. A. Sekhar
    Adviser (WR)

MINISTRY OF POWER

19. Vijay Kumar Tonk
    Deputy Director
    Central Electricity Authority

GOVERNMENT OF HARYANA

20. S.L. Aggarwal
    (Representing Engineer-in-Chief)
    Irrigation Department

GOVERNMENT OF HIMACHAL PRADESH

21. S.K. Bhatnagar
    Chief Engineer
    Irrigation and Public Health Department

INTERNATIONAL COMMISSION ON IRRIGATION AND DRAINAGE

22. M. Gopalakrishnan
    Secretary General

INTERNATIONAL COMMISSION ON LARGE DAMS

23. C.V.J. Verma
    Ex-President and Secretary General
    (Council for Power Utilities)

EXPERTS/SPECIALISTS

24. C.D. Thatte
    Former Secretary (Water Resources)

25. A.D. Mohile
    Former Chairman,CWC

26. S.C. Sud
    Chief Engineer
    (Representing INCID)

27. G.N. Mathur
    Secretary
    Central Board of Irrigation and Power

28. A.S. Chawla
    Director, CBI&P

29. Binayak Rath
    Professor of Economics
    Indian Institute of Technology, Kanpur

30. M.P. Kaushal
    Professor
    Dept. of Soil & Water Engg.
    Punjab Agricultural University
    Ludhiana

31. Paritosh C. Tyagi
    Former Chairman, CPCB

32. Krishna Pal
    RITES

33. Rema Devi
    Dept. of Civil Engineering
    IIT, Delhi

34. Somnath Mukherjee
    Ecosmart India Ltd

ANNEXURE 2

The Indus Waters Treaty, 1960

Preamble

The Government of India and the Government of Pakistan, being equally desirous of attaining the most complete and satisfactory utilisation of the waters of the Indus system of rivers and recognizing the need, therefore, of fixing and delimiting in a spirit of goodwill and friendship, the rights and obligations of each in relation to the other concerning the use of these waters and of making provision for the settlement, in a cooperative spirit, of all such questions as may hereafter arise in regard to the interpretation or application of the provisions agreed upon herein, have resolved to conclude a Treaty in furtherance of these objectives, and for this purpose have named as their plenipotentiaries:

THE GOVERNMENT OF INDIA:
Shri Jawaharlal Nehru,
Prime Minister of India,

AND

THE GOVERNMENT OF PAKISTAN:
Field Marshal Mohammad Ayub Khan, HP, H.J.,
President of Pakistan,

who, having communicated to each other their respective Full Powers and having found them in good and due form, have agreed upon the following Articles and Annexures:

ARTICLE-I

DEFINITIONS

As used in this Treaty:

(1)   The terms 'Article' and 'Annexure' mean respectively an Article of, and an Annexure to, this Treaty. Except as otherwise indicated, references to Paragraphs are to the paragraphs in the Article or in the Annexure in which the reference is made.

(2)   The term 'Tributary' of a river means any surface channel, whether in continuous or intermittent flow and by whatever name called, whose waters in the natural course would fall into that river, e.g. a tributary, a torrent, a natural drainage, an artificial drainage, a *nadi,* a *nallaha,* a *nai,* a *khad,* a *cho.* The term also includes any sub-tributary or branch or subsidiary channel, by whatever name called, whose waters, in the natural course, would directly or otherwise flow into that surface channel.

(3)   The term 'The Indus', 'The Jhelum', 'The Chenab', 'The Ravi', 'The Beas', or 'The Sutlej', means the named river (including connecting Lakes, if any) and all its Tributaries. Provided however that:
    (i)   none of the rivers named above shall be deemed to be a Tributary;
    (ii)   the Chenab shall be deemed to include the river Panjnad, and
    (iii) the river Chandra and the river Bhaga shall be deemed to be Tributaries of The Chenab.

(4)   The term 'Main' added after Indus, Jhelum, Chenab, Sutlej, Beas or Ravi means the main stem of the named river excluding its Tributaries, but including all channels and creeks of the main stem of that river and such connecting Lakes as form part of the main stem itself. The Jhelum Main shall be deemed to extend up to Verinag, and the Chenab Main up to the confluence of the river Chandra and the river Bhaga.

(5)   The term 'Eastern Rivers' means The Sutlej, The Beas and The Ravi taken together.

(6)   The term 'Western Rivers' means The Indus, the Jhelum and The Chenab taken together.

(7)   The term 'The Rivers' names all the rivers, The Sutlej, The Beas, The Ravi, The Indus, The Jhelum and The Chenab.

(8)   The term 'Connecting Lake' means any lake which receives water from, or yields water to, any of the Rivers; but any lake which occasionally and irregularly receives only the spill of any of the Rivers and returns only the whole or part of that spill is not a connecting Lake.

(9)   The term 'Agricultural Use' means the use of water for irrigation, except for irrigation of household gardens and public recreational gardens.

(10)   The term 'Domestic Use' means the use of water for:
    (a)   drinking, washing, bathing, recreation, sanitation (including the conveyance and dilution of sewage and of industrial and other wastes), stock and poultry, and other like purposes;
    (b)   household and municipal purposes (including use for household gardens and public recreational gardens); and
    (c)   industrial purposes (including mining, milling and other like purposes); but the term does not include Agricultural Use or use for the generation of hydroelectric power.

(11)   The term 'Non-Consumptive Use' means any control or use of water for navigation, floating of timber or other property, flood protection or flood control, fishing or fish culture, wild life or other like beneficial purposes,

provided that, exclusive of seepage and evaporation of water incidental to the control or use, the water (undiminished in volume within the practical range of measurement) remains in, or is returned to, the same river or its Tributaries; but the term does not include Agricultural Use or use for the generation of hydro-electric power.

(12) The term 'Transition Period' means the period beginning and ending as provided in Article II (6).

(13) The term 'Bank' means the International Bank for Reconstruction and Development.

(14) The term 'Commissioner' means either of the Commissioners appointed under the provisions of Article VIII (1) and the term 'Commission' means the Permanent Indus Commission constituted in accordance with Article VIII (3).

(15) The term 'interference with the waters' means:
 (a) Any act of withdrawal therefrom; or
 (b) Any man-made obstruction to their flow which causes a change in the volume (within the practical range of measurement) of the daily flow of the waters: provided however that an obstruction which involves only an insignificant and incidental change in the volume of the daily flow, for example, fluctuations due to afflux caused by bridge piers or a temporary bypass etc., shall not be deemed to be an interference with the waters.

(16) The term 'Effective Date' means the date on which this Treaty takes effect in accordance with the provisions of Article XII, that is, the first of April 1960.

Article-II

Provisions Regarding Eastern Rivers

(1) All the waters of the Eastern Rivers shall be available for the unrestricted use of India, except as otherwise expressly provided in this Article.

(2) Except for Domestic Use and Non-consumptive Use, Pakistan shall be under an obligation to let flow, and shall not permit any interference with, the waters of the Sutlej Main and the Ravi Main in the reaches where these rivers flow in Pakistan and have not yet finally crossed into Pakistan. The points of final crossing are the following: (a) near the new Hasta Bund upstream of Suleimanke in the case of the Sutlej Main, and (b) about one and a half miles upstream of the syphon for the B-R-B-D Link in the case of the Ravi Main.

(3) Except for Domestic Use, Non-consumptive Use and Agricultural Use (as specified in Annexure B), Pakistan shall be under an obligation to let flow, and shall not permit any interference with, the waters (while flowing in Pakistan) of any Tributary which in its natural course joins the Sutlej main or the Ravi Main before these rivers have finally crossed into Pakistan.

(4)   All the waters, while flowing in Pakistan, of any Tributary which, in its natural course, joins the Sutlej Main or the Ravi Main after these rivers have finally crossed into Pakistan shall be available for the unrestricted use of Pakistan; provided however that this provision shall not be construed as giving Pakistan any claim or right to any releases by India in any such Tributary. If Pakistan should deliver any of the waters of any such Tributary, which on the effective Date joins the Ravi Main after this river has finally crossed into Pakistan, into a reach of the Ravi Main upstream of this crossing, India shall not make use of these waters; each Party agrees to establish such discharge observation stations and make such observations as may be necessary for the determination of the component of water available for the use of Pakistan on account of the aforesaid deliveries by Pakistan, and Pakistan agrees to meet the cost of establishing the aforesaid discharge observation stations and making the aforesaid observations.

(5)   There shall be a Transition Period* during which, to the extent specified in Annexure H* India shall
   (i)   limit its withdrawals for Agricultural Use,
   (ii) limit abstractions for storages, and
   (iii) make deliveries to Pakistan from the Eastern Rivers.

(6)   The Transition Period* shall begin on 1st April 1960 and it shall end on 31st March, 1970, or, if extended under the provisions of Part 8 of Annexure H, on the date upto which it has been extended. In any event, whether or not the replacement referred to in Article IV (I) has been accomplished, the Transition Period shall end not later than 31st March, 1973.

(7)   If the Transition Period* is extended beyond 31st March, 1970, the provisions of Article V (5) shall apply.

(8)   If the Transition Period* is extended beyond 31st March 1970, the provisions of Paragraph (5) shall apply during the period of extension beyond 31st March 1970.

(9)   During the Transition Period*, Pakistan shall receive for unrestricted use the waters of the Eastern Rivers which are to be released by India in accordance with the provisions of Annexure H. After the end of the transition Period, Pakistan shall have no claim or right to release by India of any of the waters of the Eastern Rivers In case there are any releases, Pakistan shall enjoy the unrestricted use of the waters so released after they have finally crossed into Pakistan: Provided that in the event that Pakistan makes any use of these waters, Pakistan shall not acquire any right whatsoever, by prescription or otherwise, to a continuance of such release or such use.

* The transition period has ended. Provisions relating to transition period have since been implemented or have lapsed.
   * Annexure H—has not been reproduced here.

ARTICLE-III

PROVISIONS REGARDING WESTERN RIVERS

(1) Pakistan shall receive for unrestricted use all those waters of the Western Rivers which India is under obligation to let flow under the provisions of Paragraph (2).

(2) India shall be under an obligation to let flow all the waters of the Western Rivers, and shall not permit any interference with these waters except for the following uses, restricted except as provided in item (c)(ii) of Paragraph 5 of Annexure (C) in the case of the rivers, the Indus, the Jhelum and the Chenab, to the drainage basin thereof:

   (a) Domestic Use;
   (b) Non-Consumptive Use;
   (c) Agricultural Use, as set out in Annexure C; and
   (d) Generation of hydro-electric power; as set out in Annexure D.

(3) Pakistan shall have the unrestricted use of all waters originating from sources other than the Eastern Rivers which are delivered by Pakistan into The Ravi or The Sutlej, and India shall not make use of these waters. Each Party agrees to establish such discharge observation stations and make such observations as may be considered necessary by the Commission for the determination of the component of water available for the use of Pakistan on account of the aforesaid deliveries by Pakistan.

(4) Except as provided in Annexures D and E, India shall not store any water of, or construct any storage works on, the Western Rivers.

ARTICLE-IV

PROVISIONS REGARDING EASTERN RIVERS AND WESTERN RIVERS

(1) Pakistan shall use its best endeavours to construct and bring into operation, with due regard to expedition and economy, that part of a system of works which will accomplish the replacement, from the Western rivers and other sources, of water supplies for irrigation canals in Pakistan which, on 15th August 1947, were dependent on water supplies from the Eastern Rivers*.

(2) Each Party agrees that any Non-Consumptive Use made by it shall be so made as not to materially change, on account of such use, the flow in any channel to the prejudice of the uses on that channel by the other Party under the provisions of this Treaty. In executing any scheme of flood protection or flood control each party will avoid, as far as practicable, any material damage to the other Party, and any such scheme carried out by India on

---

* since implemented.

the Western Rivers shall not involve any use of water or any storage in addition to that provided under Article III.

(3) Nothing in this Treaty shall be construed as having the effect of preventing either Party from undertaking schemes of drainage, river training, conservation of soil against erosion and dredging, or from removal of stones, gravel or sand from the beds of the Rivers; provided that

    (a) in executing any of the schemes mentioned above, each Party will avoid, as far as practicable, any material damage to the other Party;

    (b) any such scheme carried out by India on the Western Rivers shall not involve any use of water or any storage in addition to that provided under Article III;

    (c) except as provided in Paragraph (5) and Article VII (1) (b), India shall not take any action to increase the catchment area, beyond the area on the Effective Date, of any natural or artificial drainage or drain which crosses into Pakistan , and shall not undertake such construction or remodeling of any drainage or drain which so crosses or falls into a drainage or drain which so crosses as might cause material damage in Pakistan or entail the construction of a new drain or enlargement of an existing drainage or drain in Pakistan; and

    (d) should Pakistan desire to increase the catchment area, beyond the area on the Effective date, of any natural or artificial drainage or drain, which receives drainage waters from India or except in an emergency, to pour any waters into it in excess of the quantities received by it as on the Effective Date, Pakistan shall, before undertaking any work for these purposes, increase the capacity of that drainage or drain to the extent necessary so as not to impair its efficacy for dealing with drainage waters received from India as on the Effective Date.

(4) Pakistan shall maintain in good order its portions of the drainages mentioned below with capacities not less that the capacities as on the Effective date:

    (i) Hudiara Drain

    (ii) Kasur Nala

    (iii) Salimshah Drain

    (iv) Fazilka Drain

(5) If India finds it necessary that any of the drainages mentioned in Paragraph (4) should be deepened or widened in Pakistan, Pakistan agrees to undertake to do so as a work of public interest, provided India agrees to pay the cost of the deepening or widening.

(6) Each Party will use its best endeavours to maintain the natural channels of the Rivers, as on the Effective Date, in such condition as will avoid, as far as practicable, any obstruction to the flow in these channels likely to cause material damage to the other party.

(7)   Neither Party will take any action which would have the effect of diverting the Ravi Main between Madhopur and Lahore, or the Sutlej Main between Harike and Suleimanke, from its natural channel between high banks.

(8)   The use of the natural channels of the Rivers for the discharge of flood or other excess waters shall be free and not subject to limitation by either Party, and neither Party shall have any claim against the other in respect of any damage caused by such use. Each Party agrees to communicate to the other Party as far in advance as practicable any information it may have in regard to such extraordinary discharges of water from reservoirs and flood flows as may affect the other Party.

(9)   Each Party declares its intention to operate its storage dams, barrages and irrigation canals in such manner, consistent with the normal operations of its hydraulic systems, as to avoid, as far as feasible, material damage to the other Party.

(10) Each Party declares its intention to prevent as far as practicable, undue pollution of the waters of the Rivers which might affect adversely uses similar in nature to those to which the waters were put on the Effective date, and agrees to take all reasonable measures to ensure that, before any sewage or industrial waste is allowed to flow into the Rivers, it will be treated, where necessary, in such manner as not materially to affect those uses: Provided that the criterion of reasonableness shall be the customary practice in similar situations on the Rivers.

(11) The Parties agree to adopt, as far as feasible, appropriate measures for the recovery, and restoration to owners, of timber and other property floated or floating down the Rivers, subject to appropriate charges being paid by the owners.

(12) The use of water for industrial purposes under Article II(2), II(3) and III(2) shall not exceed:

    (a)   in the case of an industrial process not known on the Effective Date, such quantum of use as was customary in that process on the Effective Date;

    (b)   in the case of an industrial process not known on the Effective Date:

        (i)   such quantum of use as was customary on the Effective Date in similar or in any way comparable industrial processes; or

        (ii) if there was no industrial process on the Effective Date similar or in any way comparable to the new process, such quantum of use as would not have a substantially adverse effect on the other Party.

(13) Such part of any water withdrawn for Domestic Use under the provisions of Articles II (3) and III (2) as is subsequently applied to Agricultural Use shall be accounted for as part of the Agricultural Use specified in Annexure B and Annexure C respectively; each party will use its best endeavours to return to the same river (directly or through one of its Tributaries all water withdrawn therefrom for industrial purposes and not consumed either in

the industrial processes for which it was withdrawn or in some other Domestic Use.

(14) In the event that either party should develop a use of the waters of the Rivers which is not in accordance with the provisions of this Treaty, that Party shall not acquire by reason of such use any right, by prescription or otherwise, to a continuance of such use.

(15) Except as otherwise required by the express provisions of this Treaty, nothing in this Treaty shall be construed as affecting existing territorial rights over the waters of any of the Rivers or the beds or banks thereof, as affecting existing property rights under municipal law over such waters of beds or banks.

Article—V

FINANCIAL PROVISIONS

(1) In consideration of the fact that the purpose of part of the system of works referred to in Article IV(I) is the replacement, from the Western Rivers and other sources, of water supplies for irrigation canals in Pakistan which, on 15th August, 1947, were dependent on water supplies form the Eastern Rivers, India agrees to make a fixed contribution of Pounds Sterling 62,060,000 towards the costs of these works. The amount in Pounds Sterling of this contribution shall remain unchanged irrespective of any alteration in the par value of any currency.*

(2) The sum of Pounds Sterling 62,060,000 specified in Paragraph (1) shall be paid in ten equal annual instalments on the 1st of November of each year. The first of such annual instalments shall be paid on 1st November, 1960, or if the Treaty has not entered into force by that date, then within one month after the Treaty enters into force.*

(3) Each of the instalments specified in paragraph (2) shall be paid to the Bank for the credit of the Indus Basin Development Fund to be established and administered by the Bank, and payment shall be made in Pounds Sterling, or in such other currency or currencies as may from time to time be agreed between India and the Bank.*

(4) The payments provided for under the provisions of Paragraph (3) shall be made without deduction or set-off on account of any financial claims of India on Pakistan arising otherwise than under the provisions of this Treaty: Provided that this provision shall in no way absolve Pakistan from the necessity of paying in other ways debts to India which may be outstanding against Pakistan.

(5) If, at the request of Pakistan, the Transition Period is extended** in accordance with the provision of Article II (6) and of Part 8 of Annexure H,

* since implemented.
** since lapsed.

the Bank shall thereupon pay to India out of the Indus Basin Development Fund the appropriate amount specified in the Table below:

Period of Aggregate Extension of Transition Period	Payment to India
One year	Pound Stg. 3,125,000
Two years	Pound Stg. 6,406,250
Three years	Pound Stg. 9,850,000

(6) The provisions of Article IV(1) and Article V(1) shall not be construed as conferring upon India any right to participate in the decision as to the system of works which Pakistan constructs pursuant to Article IV(1) or as constituting an assumption of any responsibility by India or as an agreement by India in regards to such works.

(7) Except for such payment as are specifically provided for in this Treaty, neither Party shall be entitled to claim any payment for observance of the provisions of this Treaty, or to make any charge for water received from it by the other Party.

ARTICLE-VI

EXCHANGE OF DATA

(1) The following data with respect to the flow in, and utilisation of the waters of, the Rivers shall be exchanged regularly between the Parties:

(a) Daily (or as observed or estimated less frequently) gauge and discharge data relating to flow of the Rivers at all observation sites.

(b) Daily extractions for or releases from reservoirs.

(c) Daily withdrawals at the heads of all canals operated by government or by a government agency (hereinafter in this Article called canals), including link canals.

(d) Daily escapes from all canals, including link canals.

(e) Daily deliveries from link canals.

These data shall be transmitted monthly by each Party to the other as soon as the data for a calendar month have been collected and tabulated, but not later than three months after the end of the month to which they relate : provided that such of the data specified above as are considered by either Party to be necessary for operational purposes shall be supplied daily or at less frequent intervals, as may be requested. Should one Party request the supply of any of these data by telegram, telephone, or wireless, it shall reimburse the other Party for the cost of transmission.

(2) If, in addition to the data specified in Paragraph (1) of this Article, either Party requests the supply of any data relating to the hydrology of the

Rivers, or to canal or reservoir operation connected with the Rivers, or to any provision of this Treaty, such data shall be supplied by the other Party to the extent that these are available.

ARTICLE-VII

FUTURE CO-OPERATION

(1) The two Parties recognize that they have a common interest in the optimum development of the Rivers, and, to that end, they declare their intention to co-operate, by mutual agreement, to the fullest possible extent. In particular:

(a) Each Party, to the extent it considers practicable and on agreement by the other Party to pay the costs to be incurred, will, at the request of the other Party, set up or install such hydro-logic observation stations within the drainage basins of the Rivers, and set up or install such meteorological observation stations relating thereto and carry out such observations there at, as may be requested, and will supply the data so obtained.

(b) Each Party, to the extent it considers practicable and on agreement by the other Party to pay the costs to be incurred, will, at the request of the other party, carry out such new drainage works as may be required in connection with new drainage works of the other Party.

(c) At the request of either Party, the two Parties may, by mutual agreement co-operate in under-taking engineering works in the Rivers.

The formal arrangements, in each case shall be as agreed upon between the Parties.

(2) If either Party plans to construct any engineering works which would cause interference with the waters of any of the Rivers and which in its opinion, would affect the other Party materially, it shall notify the other Party of its plans and shall supply such data relating to the work as may be available and as would enable the other Party to inform itself of the nature, magnitude and effect of the work. If a work would cause interference with the waters of any of the Rivers but would not, in the opinion of the Party planning it, affect the other Party materially, nevertheless the Party planning the work shall, on request, supply the other Party with such data regarding the nature, magnitude and effect, if any, of the work as may be available.

ARTICLE-VIII

PERMANENT INDUS COMMISSION

(1) India and Pakistan shall each create a permanent post of Commissioner for Indus Waters and shall appoint to this post, as often as a vacancy occurs, a person who should ordinarily be a high-ranking engineer competent in the field of hydrology and water-use. Unless either Government should

decide to take up any particular question directly with the other Government, each Commissioner will be representative of his Government for all matters arising out of this Treaty, and will serve as the regular channel of communication on all matters relating to the implementation of the Treaty, and in particular, with respect to;

    (a)   The furnishing or exchange of information or data provided for in the Treaty, and

    (b)   The giving of any notice or response to any notice provided for in the Treaty.

(2)   The status of each Commissioner and his duties and responsibilities towards his Government will be determined by that Government.

(3)   The two Commissioners shall together form the Permanent Indus Commission.

(4)   The purpose and functions of the Commission shall be able to establish and maintain co-operative arrangements for the implementation of this Treaty, to promote co-operation between the Parties in the development of the waters of the Rivers and, in particular,

    (a)   to study and report to the two Governments on any problem relating to the development of the waters of the Rivers which may be jointly referred to the Commission by the two Governments : in the event that a reference is made by the one government alone, the Commissioner of the other Government shall obtain the authorisation of his Government before he proceeds to act on the reference;

    (b)   to make every effort to settle promptly, in accordance with the provisions of Article IX (1), any question arising thereunder;

    (c)   to undertake, once in every five years, a general tour of inspection of the Rivers for ascertaining the facts connected with various developments and works on the Rivers.

    (d)   To undertake promptly, at the request of either Commissioner, a tour of inspection of such works or sites on the Rivers as may be considered necessary by him for ascertaining the facts connected with those works or sites; and

    (e)   To take, during the Transition Period, such steps as may be necessary for the implementation of the Provisions of Annexure H.

(5)   The Commission shall meet regularly at least once a year, alternately in India and Pakistan. This regular annual meeting shall be held in November or in such other month as may be agreed upon between the Commissioners. The Commission shall also meet when requested by either Commissioner.

(6)   To enable the Commissioners to perform their functions in the Commission, each Government agrees to accord to the Commissioner of the other Government the same privileges and immunities as are accorded to representatives of member States to the Principal and subsidiary organs of the United nations under Sections 11, 12 and 13 of Article IV of the Convention on the Privileges and immunities of the United Nations (dated

13th February, 1946) during the periods specified in those sections. It is understood and agreed that these privileges and immunities are accorded to the Commissioners not for the personal benefit of the individuals themselves but in order to safeguard the independent exercise of their functions in connection with the Commission; consequently, the Government appointing the Commissioner not only has the right but is under a duty to waive the immunity of its Commissioners in any case where, in the opinion of the appointing Government, the immunity would impede the course of justice and can be waived without prejudice to the purpose for which the immunity is accorded.

(7) For the purpose of the inspections specified in Paragraph (4) (c) and (d), each Commissioner may be accompanied by two Advisers or assistants to whom appropriate facilities will be accorded.

(8) The Commission shall submit to the Government of India and to the Government of Pakistan, before the first of June of every year, a report on its work for the year ended on the preceding 31st of March, and may submit to the two Governments other reports at such times as it may think desirable.

(9) Each Government shall bear the expenses of its Commissioner and his ordinary staff. The cost of any special staff required in connection with the work mentioned in Article VII (1) shall be borne as provided therein.

(10) The Commission shall determine its own procedures.

ARTICLE-XI

SETTLEMENT OF DIFFERENCES AND DISPUTES

(1) Any question which arises between the Parties concerning the interpretation or application of this Treaty or the existence of any fact which, if established, might constitute a breach of this Treaty shall first be examined by the Commission, which will endeavour to resolve the question by agreement.

(2) If the Commission does not reach agreement on any of the questions mentioned in Paragraph (1), then a difference will be deemed to have arisen, which shall be dealt with as follows:

    (a) Any difference which, in the opinion of either Commissioner, falls within the provisions of Part 1 of Annexure F* shall, at the request of either Commissioner, be dealt with by a Neutral Expert in accordance with the provisions of Part 2 of Annexure F.*

    (b) If the difference does not come within the provisions of Paragraph (2) (a), or if a Neutral Expert, in accordance with the provisions of Paragraph 7 of Annexure F,* has informed the Commission that, in his opinion the difference, or a part thereof, should be treated as a

---

* Annexure F—has not been reproduced here.

dispute, then dispute will be deemed to have arisen which shall be settled in accordance with the provisions of Paragraph (3), (4) and (5). Provided that, at the discretion of the Commission, any difference may either be dealt with by a Neutral Expert in accordance with the provisions of part 2 of Annexure F* or be deemed to be dispute to be settled in accordance with the provisions of Paragraph (3), (4), and (5), or may be settled in any other way agreed upon by the Commission.

(3) As soon as a dispute to be settled in accordance with this and the succeeding paragraphs of this Article has arisen, the Commission shall, at the request of either Commissioner, report the fact to the two Governments as early as practicable, stating in its report the points on which the Commission is in agreement and the issues in dispute, the views of each Commissioner on these issues and his reasons therefore.

(4) Either Government may, following receipt of the report referred to in paragraph (3), or if it comes to the conclusion that this report is being unduly delayed in the Commission, invite the other Government to resolve the dispute by agreement. In doing so it shall state the names of its negotiators and their readiness to meet with the negotiators to be appointed by the other Government at a time and place to be indicated by the other Government. To assist in these negotiations, the two Governments may agree to enlist the services of one or more mediators acceptable to them.

(5) A court of Arbitration shall be established to resolve the dispute in the manner provided by Annexure G*.

   (a) upon agreement between the Parties to do so; or

   (b) at the request of either Party, if, after negotiations have begun pursuant to Paragraph (4), in its opinion the dispute is not likely to be resolved by negotiation or mediation; or

   (c) at the request of either Party, if, after the expiry of one month following receipt by the other Government of the invitation referred to in Paragraph (4), that Party comes to the conclusion that the other Government is unduly delaying the negotiations.

(6) The provisions of Paragraphs (3), (4) and (5) shall not apply to any difference while it is being dealt with by a Neutral Expert.

ARTICLE-X

EMERGENCY PROVISION*

If, at any time prior to 31st March 1965, Pakistan should represent to the Bank that, because of the outbreak of large-scale international hostilities arising out of causes beyond the control of Pakistan, it is unable to obtain from abroad the materials and equipment necessary for the completion, by 31st March, 1973, of that part of the

* Annexure G—has not been reproduced here.

* Lapsed on 31st March 1965.

system of works referred to in Article IV(1) which related to the replacement referred to therein, (hereinafter referred to as the 'replacement element') and if, after consideration of this representation in consultation with India, the Bank is of the opinion that

(a)   these hostilities are on the scale of which the consequence is that Pakistan is unable to obtain in time such materials and equipment as must be procured from abroad for the completion, by 31st March 1973, of the replacement element, and

(b)   since the Effective Date, Pakistan has taken all reasonable steps to obtain the said materials and equipment and, with such resources of materials and equipment as have been available to Pakistan both from within Pakistan and from abroad, has carried forward the construction of the replacement element with due diligence and all reasonable expedition, the Bank shall immediately notify each of the Parties accordingly. The Parties undertake, without prejudice to the provisions of Article XII (3) and (4), that, on being so notified, they will forthwith consult together and enlist the good offices of the Bank in their consultations, with a view to reaching mutual agreement as to whether or not, in the light of all the circumstances then prevailing, any modifications of the provisions of the Treaty are appropriate and advisable and, if so, the nature and the extent of the modifications.

ARTICLE-XI

GENERAL PROVISIONS

(1)   It is expressly understood that

(a)   this Treaty governs the rights and obligations of each party in relation to the other with respect only to the use of the waters of the Rivers and matters incidental thereto; and

(b)   nothing contained in this Treaty, and nothing arising out of the execution thereof, shall be construed as constituting a recognition or waiver (whether tacit, by implication or otherwise) of any rights or claims whatsoever of either of the Parties other than those rights or claims which are expressly recognized or waived in this Treaty.

Each of the Parties agrees that it will not invoke this Treaty, anything contained therein, or anything arising out of the execution thereof, in support of any of its own rights or claims whatsoever or in disputing any of the rights or claims whatsoever of the other Party, other than those rights or claims which are expressly recognised or waived in this Treaty.

(2)   Nothing in this Treaty shall be construed by the Parties as in any way establishing any general principle of law or any precedent.

(3)   The rights and obligations of each party under this Treaty shall remain unaffected by any provisions contained in, or by anything arising out of the execution of any agreement establishing the Indus Basin Development Fund.

Article-XII

FINAL PROVISIONS

(1) This Treaty consists of the Preamble, the Articles hereof and Annexures A to H hereto, and may be cited as 'The Indus Waters Treaty 1960'.

(2) This Treaty shall be ratified and the ratifications thereof shall be exchanged in New Delhi. It shall enter into force upto the exchange of ratifications, and will then take effect retrospectively from the first of April 1960.

(3) The provisions of this Treaty may from time to time be modified by a duly ratified treaty concluded for that purpose between the two Governments.

(4) The provisions of this Treaty, or the provisions of this Treaty as modified under the provisions of Paragraph (3), shall continue in force until terminated by a duly ratified treaty concluded for that purpose between the two Governments.

IN WITNESS WHEREOF the respective Plenipotentiaries have signed this Treaty and have hereunto affixed their seals.

Done in triplicate in English at Karachi on this Nineteenth day of September, 1960.

FOR THE GOVERNMENT OF INDIA :
(Sd.) Jawaharlal Nehru ....

FOR THE GOVERNMENT OF PAKISTAN :
(Sd.) Mohammed Ayub Khan ....
Field Marshal, H.P., H.J.

FOR THE INTERNATIONAL BANK FOR RECONSTRUCTION AND DEVEL-OPMENT for the purposes specified in Articles V and X and Annexures F, G and H:
(Sd.) W.A. B. Iliff ....

ANNEXURE A—EXCHANGE OF NOTES BETWEEN GOVERNMENT OF INDIA AND GOVERNMENT OF PAKISTAN

ANNEXURE B—AGRICULTURAL USE BY PAKISTAN FROM CERTAIN TRIBUTARIES OF THE RAVI

ANNEXURE C—AGRICULTURAL USE BY INDIA FROM THE WESTERN RIVERS

ANNEXURE D—GENERATION OF HYDRO-ELECTRIC POWER BY INDIA ON THE WESTERN RIVERS

ANNEXURE E—STORAGE OF WATERS BY INDIA ON THE WESTERN RIVERS

ANNEXURE F—NEUTRAL EXPERT

ANNEXURE G—COURT OF ARBITRATION

ANNEXURE H—TRANSITIONAL ARRANGEMENTS

(The text of these annexures has not been reproduced here.)

ANNEXURE 3

Inter-State Agreements on Ravi-Beas Waters

A. 1955 Agreement/Decision

Record of decisions arrived at the inter-state conference on the Development and Utilisation of the Water of the Rivers Ravi and Beas, held in Room No.12, North Block, New Delhi on the 29th January 1955.

After a brief discussion of the demands for the waters as given by the various States, the following decisions were taken:

1. The supplies both flow and storage in the rivers Ravi and Beas over and above the actual pre-partition use based on the mean supplies in the rivers, shall be allocated as under:-

Share of Punjab	–	5.90 MAF
Share of Kashmir	–	0.65 MAF
Share of Rajasthan	–	8.00 MAF
Share of PEPSU	–	1.30 MAF
Total	–	15.85 MAF

In case of any variation in total supplies, the shares shall be changed pro-rata on the above allocations subject to the conditions that no change shall be made in the allocation for Kashmir State which shall remain as 0.65 MAF.

2. The distribution of flow supplies shall be in the same ratio as that of allocation mentioned above.

3. The splitting up of the allocated supplies between Kharif and Rabi may be left to Engineers, the matter may be referred to the Government of India if they can not arrive at an agreement of this issue.

4. The proposed capacity of Madhopur-Beas Link may be increased from 8000 to 10,000 cusecs.

5. The question of allocation of the cost of water including the cost of storage and other works may be taken up separately as the conference was concerned only with the distribution of supplies.

6. It is left to each State to decide as to how best to utilise the supplies allocated to it. The States, however, must submit their proposals in this regard immediately to the Government of India (Planning Commission).

B.   Relevant extracts from the Govt. of India Notification dated 24th March 1976 on sharing of Ravi-Beas Waters between Punjab and Haryana arising out of reorganisation of the state of Punjab.

GOVERNMENT OF INDIA
(Bharat Sarkar)
Ministry of Agriculture and Irrigation
(Krishi Aur Sinchai Mantralaya)
(Department of Irrigation)
(Sinchai Vibhag)

New Delhi, the 24 April, 1976

NOTIFICATION

....
....
....
....

NOW, THEREFORE, in exercise of the powers conferred by sub-section (1) of section 78 of the Punjab Re-organisation Act 1966 (31 of 1966), the Central Government hereby makes the following determination, namely:–

....
....
....
.... the Central Government hereby directs that out of the water which would have become available to the erstwhile state of Punjab on completion of the Beas Project (0.12 MAF whereof is earmarked for Delhi water supply), the State of Haryana will get 3.5 MAF and the state of Punjab will get the remaining quantity not exceeding 3.5 MAF. When further conservation works on the Ravi are completed, Punjab will get 3.5 of 7.2 MAF which is the share of the erstwhile State of Punjab. The remaining 0.08 MAF, out of 7.2 MAF is recommended as additional quantum of water for Delhi water supply for acceptance by both the Government of Punjab and Haryana.

....
....
....
....

C. 1981 AGREEMENT

WHEREAS under the Indus Waters Treaty of 1960, the waters of the three rivers, namely Sutlej, Beas and Ravi became available for unrestricted use by India after 31st March, 1970; and

WHEREAS while at the time of signing of the said treaty, the waters of the Sutlej had already been planned to be utilised for the Bhakra Nangal Project, the surplus flow of rivers Ravi and Beas over and above the pre-partition use, was allocated by agreement, in 1955 (hereinafter called the 1955 Agreement), between the concerned States as follows, namely:

Punjab	–	7.20 MAF (including 1.30 MAF for PEPSU)
Rajasthan	–	8.00 MAF
Jammu and Kashmir	–	0.65 MAF
		15.85 MAF

and, for the purpose of the said allocation, the availability of water was based on the flow series of the said rivers for the year 1921-1945; and

WHEREAS the Central Government issued a notification on 24th March, 1976, allocating 3.50 MAF of the waters becoming available as a result of Beas Project to Haryana and the balance not exceeding 3.50 MAF to Punjab out of the total surplus Ravi-Beas waters of 7.20 MAF falling to the share of erstwhile State of Punjab after setting aside 0.20 MAF for Delhi drinking water supply; and

WHEREAS the Punjab Government sought a review of the aforesaid notification for increasing the allocation of Punjab and linked this matter to the construction of the Sutlej-Yamuna Link Canal for Haryana in Punjab territory; and

WHEREAS the Government of Haryana filed a suit in the Supreme Court praying inter alia that a directive be issued to Punjab for expeditiously undertaking construction of the Sutlej-Yamuna Link Canal in Punjab territory and for declaring that the notification of the Government of India allocating the waters becoming available as a result of the Beas Project issued on 24th March, 1976, is final and binding; and

WHEREAS the Punjab Government also filed a suit in the Supreme Court challenging the competence of the Central Government to enact section 78 of the Punjab Re-organisation Act, 1966 and notwithstanding this, questioning the notification issued under section 78 of the said Act; and

WHEREAS adjournment has been sought from time to time in hearing of the suits filed in the Supreme Court by Haryana and Punjab to enable the parties to arrive at a mutually acceptable settlement of the differences that have arisen; and

WHEREAS discussions have been held by the Prime Minister of India and Union Minister of Law, Justice and Company Affairs with the Chief Ministers of Haryana, Punjab and Rajasthan.

Now, therefore, we the Chief Ministers of Haryana, Rajasthan and Punjab keeping in view the overall national interest and desirous of speedy and optimum utilisation of waters of the Ravi and Beas Rivers and also having regard to the imperative need to resolve speedily the difference relating to the use of these waters in a spirit of give and take do hereby agree as under:–

(i) According to the flow series 1921–60, the total mean Supply of Ravi-Beas waters is 20.56 MAF. Deducting the pre-partition uses of 3.13 MAF and transit losses in the Madhopur Beas Link of 0.26 MAF the net surplus Ravi-Beas waters according to the flow series 1921–60 is 17.17 MAF as against the corresponding figures of 15.85 MAF for the flow series 1921–45, which forms the basis of water allocation under the 1955 Agreement. It is now hereby agreed that the mean supply of 17.17 MAF (flow and storage) may be reallocated as under:–

Share of Punjab	–	4.22 MAF
Share of Haryana	–	3.50 MAF
Share of Rajasthan	–	8.60 MAF
Quantity earmarked for Delhi water supply	–	0.20 MAF
Share of J&K	–	0.65 MAF
		17.17 MAF

In case of any variation in the figure of 17.17 MAF in any year, the shares shall be changed pro-rata of the above revised allocations subject to the condition that no change shall be made in the allocation of Jammu and Kashmir which shall remain fixed as 0.65 MAF as stipulated in the 1955 Agreement. The quantity of 0.20 MAF for Delhi water supply stands as already allocated.

(ii) Until such time as Rajasthan is in a position to utilise its full share, Punjab shall be free to utilise the water surplus to Rajasthan's requirements. As Rajasthan will soon be able to utlise its share Punjab shall make adequate alternative arrangements expeditiously for irrigation of its own lands by the time Rajasthan is in a position to utilise its full share. As a result, it is expected that during this transitional period when Rajasthan's require-ments would not exceed 8.0 MAF, 4.82 MAF of water should be available to Punjab in a mean year when the availability is 17.17 MAF.

(iii) The Bhakra and Beas Management Board (BBMB) shall be permitted to take all necessary measures for carrying out measurements and for ensuring delivery of supplies to all the concerned States in accordance with their entitlements such as rating the gauge discharge curves, instal-lation of self-recording gauges, taking observations without any hindrance of the discharge measurements. The Selection of the control points at which the Bhakra and Beas Management Board would take appropriate measures as mentioned above shall include, but be not limited to all points at which Bhakra and/or Ravi/Beas discharges are being shared by more than one State and all regulation points on the concerned rivers and Canals for determining the shareable supplies. The decision of the Bhakra Beas Management Board would be binding in so far as the selection of the control points is concerned for the purposes of taking discharge measurements to facilitate equitable distribution of the waters but if any State Government contests the decision, the Central Government shall decide the matter within 3 months and this decision shall be final and binding. All the concerned State Governments shall cooperate fully and shall promptly carry out day-to-day directions of the Bhakra Beas Management Board in regard to regulation and control of supplies, operation of gates and any other matters, in their territories, for ensuring delivery of supplies as determined by Bhakra Beas Management Board in accordance with their entitlements as provided under the Agreement.

(iv)   The Sutlej-Yamuna Link Canal Project shall be implemented in a time bound manner so far as the canal and appurtenant works in the Punjab territory are concerned within a maximum period of two years from the date of signing of the Agreement so that Haryana is enabled to draw its allocated share of waters. The canal capacity for the purpose of design of the canal shall be mutually agreed upon between Punjab and Haryana within 15 days, failing which it shall be 6500 cusecs, as recommended by former Chairman, Central Water Commission.

Regarding the claim of Rajasthan to convey 0.57 MAF of waters through Sutlej-Yamuna Link-Bhakra system, Secretary, Ministry of Irrigation, Government of India will hold discussions with Punjab, Haryana and Rajasthan with a view to reaching an acceptable solution. These discussions shall be concluded in a period of 15 days from the date of affixing signatures herein and before the work starts. If no mutually acceptable agreement is reached, the decision of Secretary, Ministry of Irrigation to be given within this period shall be binding on all the parties. In case it is found necessary to increase the capacity of Sutlej-Yamuna Link Canal beyond that decided under above sub-para in any or entire reach thereof, the states concerned shall implement the link canal in a time bound manner with such increased capacity at the cost of Rajasthan Government.

The differences with regard to the alignment of the Link Canal and appurtenant works in the Punjab territory would be discussed by the Haryana and Punjab Governments who should agree to a mutually acceptable canal alignment in Punjab territory including appurtenant works within a period of three months from the date of signing of this Agreement. If however, the State Governments are unable to reach complete agreement within this period, the matter shall be decided by the Central Government within a period of two weeks. Both the State Governments shall cooperate fully to enable Central Govt. to take timely decision in this regard. The decisions of the Central Government in this matter shall be final and binding on both the Governments and the canal and appurtenant works in Punjab territory shall be implemented in full by Punjab Government. However, work on the already agreed reaches of the alignment would start within fifteen days of the signing of the agreement and work within the other reaches immediately after the alignment has been decided. Haryana shall provided necessary funds to the Punjab Governments for surveys, investigations and constructions of the Link canal and appurtenant works in Punjab territory, where as a result of acquisition of land, extreme hardship is caused to families, the Punjab Government shall forward to the Haryana Government suitable proposals for relieving such hardship in line with such schemes in Punjab undertaken in respect of similar Canal woks in Punjab territory. The Haryana Government shall arrange to bear the cost of such proposals. In the event, however, of any difference of opinion arising on the question of sharing such cost, the parties shall abide by decision of the Secretary, Ministry of Irrigation, Government of India. The progress of the work shall not, however, be delayed on this account. The Central Government will be requested to monitor the progress of the work being carried out in Punjab territory.

The Agreement reached in paras (i) to (iv) above shall be implemented in full by the Government of Haryana, Rajasthan and Punjab. If any signatory States feels that any of the provisions of the Agreement are not being complied with, the matter shall be referred to the Central Government whose decisions shall be binding on all the States. In this respect the Central Government shall be competent to issue such directions or take such measures as may be appropriate to the circumstances of the case to facilitate and ensure such compliance.

The suits filed by the Governments of Haryana and Punjab in the Supreme Court would be withdrawn by the respective Governments without any reservations whatsoever but subject to the terms of this Agreement.

The Notifications of the Government of India allocating the waters becoming available as a result of the Beas Project issued on 24th March, 1976, and published in the Gazette of India, Part II, Section-3, Sub Section (ii) as well as the 1955 Agreement stand modified to the extent varied by this Agreement and shall be deemed to be in force as modified herein.

In case of any difference on interpretation of this Agreement, the matter will be referred to the Central Government whose decision shall be final.

We place on record and gratefully acknowledge the assistance and advice given by our respected Prime Minister Smt.Indira Gandhi in arriving at this expeditious and amicable settlement.

NEW DELHI, the 31st December. 1981.

Sd/-	Sd/-	Sd/-
(BHAJAN LAL)	(SHIV CHARAN MATHUR)	(DARBARA SINGH)
Chief Minister of	Chief Minister of	Chief Minister of
Haryana	Rajasthan	Punjab

In presence of:

Sd/-
(Indira Gandhi)
PRIME MINISTER OF INDIA

ANNEXURE 4

Conversion Factors

Length

1 inch	25.4 mm = 2.54 cm
1 cm	0.394 inch
1 metre	3.2808 ft.
1 ft.	30.48 cm
1 km	0.621 mile
1 mile	1.61 km

Area

1 sq metre	10.764 ft^2
1 sq.ft.	0.0093 m^2
1 ha = 10,000 m^2	2.4711 acares
1 acre	0.4047 ha
1 km^2	100 ha = 0.386 mile
1 mile^2	2.59 km^2 = 259 ha = 640 acre

Volume

1 m^3	35.315 ft^3 = 1 kilo litre = 1000 litre
1 ft^3	0.0283 m^2 = 28.32 litres = 6.23 UK gallons
1 Ac ft	1233.48 m^3
1 m^3	0.00081 Acft.
1 ha m	8.10 Acre ft = 10,000 m^3
1 Acre feet	0.1233 ha m = 43,560 ft^3
1 gallon	4.546 litres
1 Mcm	810.71 Acre ft.
1 MAF	1233.48 Mm^3
1 Mm^3	35.3147 M ft^3
1 TMC	28.317 .M.m^3 = 22956.87 acre ft.
1 Mm^3	0.0353 TMC
1 milliard	1000 million

| 1 Km³ | 1 BCM = 0.81 Macft = 10^9m³ = 1 milliard m³ |
| 1 MAF | 1.233 km³ = 43.56 TMC |

Rates of flow

1 m³/sec	35.3147 cusec = 1000 litres/sec
1 cubic feet/sec per day	1.9835 Acre feet per day
1 MGD	1.8581 ft³/sec
1 Ha.m/100 sq. km	20.996 acrefeet/100 sq. miles

Weight

| 1 Ton | 1.01605 Tonne (metric) |
| 1 maund | 82.2857 pounds =37.324 kg. |

General

1 lakh	100,000
1 crore	10 million
1 rupee	100 paise

BIBLIOGRAPHY

I. Official Publications

INDIA

GOVERNMENT OF INDIA

Report of the Irrigation Commission (1903).
Partition Secretariat, Partition Proceedings, vol. VI, April 1950.
Gazette Extraordinary, 1947.

MINISTRY OF FINANCE, ECONOMIC DIVISION

Economic Survey (2001–2), 2002.
Economic Survey (2002–3), 2003.

MINISTRY OF FOOD AND AGRICULTURE

Sen, S.R., *Growth and Instability in Indian Agriculture: Agricultural situation in India* (1967).

MINISTRY OF HUMAN RESOURCE DEVELOPMENT

Department of Women and Child Development, Annual Report (2002–3).

MINISTRY OF INFORMATION AND BROADCASTING

Publication Division, *Speeches of Jawaharlal Nehru*.

MINISTRY OF IRRIGATION AND POWER

Report of the Irrigation Commission (1972).

MINISTRY OF IRRIGATION AND AGRICULTURE

Report of the committee of technical experts regarding causes of breaches in River Sutlej and permanent solution of problem, November 1978.

MINISTRY OF PLANNING

Planning Commission, *First Five Year Plan* (1952).

————— *Third Five Year Plan* (1961).
————— *Seventh Five Year Plan* (1985–90).
————— *Tenth Five Year Plan* (2002–7).

MINISTRY OF POWER

Bhakra Beas Management Board, (various years) Annual Administration Reports.
————— (1988), *History of Bhakra Nangal Project.*

MINISTRY OF WATER RESOURCES

Waterlogging, soil salinity and alkalinity, Report of the working group on problem identification in irrigated areas with suggested remedial measures (1991), New Delhi, 1992.
Report of the National Commission for Integrated Water Resources Development (1999).
Report of the Working Group on Flood Control Programme for the Tenth Five Year Plan (2002–7), New Delhi, 2001.
Report of the Experts Committee on the Implementation of the Recommendations of the National Flood Commission (2003), (Rangachari Committee).

GOVERNMENT OF HARYANA

Statistical Abstracts of Haryana (various years).

GOVERNMENT OF PUNJAB

Bhakra Nangal Project Report (1955), Public Works Department,
Statistical Abstracts of Punjab (various years), Simla.

GOVERNMENT OF RAJASTHAN

District Gazetteer of Ganganagar (1972), Director.

UNITED KINGDOM

GOVERNMENT OF UK

Her Majesty's Stationary Office, London.
Constitutional Relations between Britain and India.
The Transfer of Power 1942–7, vol. XII.
Nicholas Mansergh (Editor in chief), Penderel Moon (Editor), 1983.

UNITED STATES OF AMERICA

GOVERNMENT OF USA

————— Printing office, Report on Land and Water Development in the Indus plain (1964).

II. Other References

Baxter, R.R. (1967), see Garretson, *et al.*

Bhalla, G.S. and Chadha (1983), *Green Revolution and the Small Peasant,* Concept Publishing Company, New Delhi.

Bhatia, Ramesh and R.P.S. Malik (2003), *Multiplier effects of dams, A case study of the Bhakra Multi-purpose Dam in India, Draft Report.*

Bhatia, Ramesh, Monica Scatasta, and Rita Cestti (eds) (forthcoming), *Indirect Economic Impacts of Dams,* World Bank, Washington DC, USA.

Birla Institute of Scientific Research (BISR) (1981), *Agricultural Growth and Employment Shifts in Punjab,* New Delhi.

Blyn, George (1966), *Agricultural trends in India, 1891 to 1947: Output availability and Productivity,* University of Pennsylvania Press, Philadelphia, USA.

Buckley, R.B. (1893), *Irrigation Works in India and Egypt,* Encyclopaedia Britannica, London, 1970.

Central Board of Irrigation and Power (1965), *Development of Irrigation in India,* New Delhi.

Centre for Monitoring Indian Economy (CMIE) (2002), *Economic Intelligence Service-Agriculture,* Mumbai.

Chopra, Kanchan Ratna (1972), *Dualism and Investment Patterns,* Tata McGraw-Hill Publishing Co., New Delhi.

Chopra, R.N. (1986), *Green Revolution in India,* Intellectual Publishing House, Delhi.

Council of Power Utilities (2000), *Profile of Power Utilities and Non-utilities in India,* New Delhi.

Dantwala, M.L. (1996), Dilemmas of Growth: The Indian Experience (ed.), by Pravin Visaria *et al.*, © Indian Society of Agricultural Economics, Sage Publications, New Delhi.

Darling, Malcolm L. (1947), *Punjab Peasant in Prosperity and Debt,* Oxford University Press, London.

Dev, D.S. *et al.* (1993), *Rural Industrialisation, Lessons from a Case Study,* Ludhiana.

Dhawan, B.D. (1993), Institute of Economic Growth, *Trends and Tendencies in Indian Irrigated Agriculture,* Commonwealth Publishers, New Delhi.

Garretson, A.H. *et al.* (ed.) (1967), *The Law of International Drainage Basins,* Institute of International Law, New York.

Gulhati, N.D. (1968), 'The Mountains and Rivers of India', 21st International Geographical Congress India (ed.) B.C. Law, Chapter XX—*The Indus and its Tributaries.*

——— (1973), *Indus Waters Treaty: An exercise in international mediation,* Allied Publishers, New Delhi.

Handa, C.L. and O.P. Chadha (1953), 'Bhakra Dam Salient Design Features', *Indian Journal of River Valley Development.*

Rao, Hanumantha, C.H. (1994), Institute of Economic Growth, *Agricultural Growth, Rural Poverty and Environmental Degradation in India*, Oxford University Press, New Delhi.

Hazell, Peter and R. Slade (1982), *Muda Dam Project in Malaysia* or Hazell P. and C. Ramasamy, *Impacts of Green Revolution in South India* (1991), John Hopkins University Press, Baltimore, Maryland.

Jeffery, R. (1974), *Modern Asian studies*.

Kalhana's, *Rajatarangini* (1900) (M.A. Stein's translation and commentaries), Motilal Banarasidass Publishers, Delhi. (Reprinted 1989)

Lilienthal (1966), *The Journals of David E. Lilienthal*; vol. III, Harper & Row, New York, 1996.

Mccully, Patrick (1996), *Silenced Rivers: The Ecology and Politics of Large Dams*, Zed Books, London.

Menon, V.P. (1968), *The Transfer of Power in India*, Orient Longman, Bombay.

Minhas, B.S. and T.N. Srinivasan (1966), *Agricultural Strategy Analysis, Yojana*.

Minhas, B.S., K.S. Parikh, and T.N. Srinivasan with S.A. Marglin and T.E. Weissko (1972), *Scheduling the operation of the Bhakra System*, Statistical Publishing Society, Calcutta.

Moon, Penderel (1998), *Divide and Quit*, Oxford University Press, New Delhi.

Paddock, William and Paul, *Famine 1975! America's Decision: Who will survive?* Little Brown, Boston, 1967.

Raj, K.N. (1960), *Economic aspects of the Bhakra Nangal Project*, Asia Publishing House, New Delhi.

Randhawa, M.S. (1986), *A History of Agriculture in India*, Indian Council of Agricultural Research, New Delhi.

Rangachari, R. (2001), *The Study on DVC reservoirs, The role of storage dams in the management of floods*. Ninth National Water Convention.

Rangachari, R. *et al.* (2000) (mimeo), *Large Dams: India's experience*, A Report prepared for the World Commission on Dams.

Razavi, Shahra (ed.) (2000), *Gendered Poverty and Well Being*, © Institute of Social Studies, The Hague, Blackwell Publishers, Oxford, UK.

Roy, Arundhati (1999), 'The Greater Common Good', *Outlook*, New Delhi.

Sain, Kanwar (1978), *Reminiscences of an Engineer*, Young Asia Publications, New Delhi.

Saith, R. and B. Harris-White, See Razavi, Shahra (2000).

Sarma, J.S. (1982), *Agricultural Policy in India: Growth with Equity*, International Development Research Centre, Canada.

Singh, Himmat (2001), *Green Revolutions reconsidered-The rural world of contemporary Punjab*, Oxford University Press, New Delhi.

Singh, Jagman (1998), *My trysts with the projects-Bhakra and Beas*, Uppal Publishing House, New Delhi.

Sudha, S. and S. Irudaya Rajan (1981–91), *Female Demographic Disadvantage in India, Sex selective abortions and female infanticide*—Working Paper No. 288 of the Centre for Development Studies, Thiruvananthapuram, 1997.

Seervai, H.M. (1994), *Partition of India: Legend and Reality*, N.M. Tripathy Publishers, Bombay (Second edition).

Shiva, Vandana (1987), *Violence and Natural Resource Conflict: A Case Study of Punjab*, Report for UN University, Tokyo.

———— (1988), *Staying Alive, Women, Ecology and Development in India*, Kali for Women, New Delhi.

Spate, O.H.K. (1954), *India and Pakistan: A General and Regional Geography*, London.

Subramaniam, C. (1979), *The New Strategy in Indian agriculture*, Vikas Publishing House, New Delhi.

Uppal, H.L., CBIP (1972), *Serious Water Logging in Punjab and Haryana, How Cured and Measures to Prevent its Recurrence*, New Delhi.

Vaidyanathan, A. (1999), *Water Resource Management: Institutions and Irrigation Development in India*, Oxford University Press, New Delhi.

Venkateswarlu, B. (1985), *Dynamics of Green Revolution in India*, Agricole Publishing Academy, New Delhi.

Verghese, B.G. (1994), *Winning the Future*, Konark Publishers Pvt. Ltd., Delhi.

WCD November (2000), *Dams and Development*, The Report of the World Commission on Dams, Earthscan Publications.

World Bank (2003), *World Development Indicators*, Washington DC, USA.

Ziegler, Philip (1985), *Mountbatten*, Alfred A. Knopf Inc., Reprinted by Harper & Row, New York.

INDEX

afforestation 190–1
Agarwal, Bina 119
Agricultural Commission 91, 92
agriculture(al) 104, 201–3
  new strategies for 5–6
  policy 80
  technologies 202
Alagh, Yogender K. 211
Ali, Mohammad 87
All India Soil and Land Use Survey
  Organization 190
anti-dam lobby 3
Arbitral Award 27
Attlee, Clement 20
Auden, W.H. 24

Baspa Hydroelectric Project 189
Beas river 11, 12, 26, 30–2, 34
Beas-Sutlej Link Project 33–4, 139
Bhakra Control Board (BCB) 67,
  68, 165
Bhakra dam 6, 17–20, 26–7, 51, 58,
  60–2, 213, 216
  construction of 58–9
  and displacement, resettlement
    and rehabilitation of 62–7
  irrigation from 19
  land acquired for 62–3
  Organization 155, 186
  powerhouses 138–43
Bhakra reservoir 9, 69
  effect of, on high flows 153–5
  and flood control 149–51, 162–3

sedimentation in 183–91
transmission network 147
Bhakra and Beas Management Board
  (BBMB) 7, 68–70, 87–8, 145,
  160, 161, 190
  on flood control 162
  flood forecast by 158
Bhakra-Nangal Agreement 1959 69
Bhakra-Nangal canal system 82–91
  development of irrigation under
    8, 88
  opening of 86–7
Bhakra-Nangal power plants 61, 136
Bhakra-Nangal Project 6–7, 26,
  56–79, 131
  costs and benefits of 75–6
  effects on agriculture 95
  hydro generating units under 141
  impact of 192–205
  incidental benefits from 165–78
  indirect benefits from 178–82
  irrigation from 9, 75, 80–4
  migrant labour 181–2
  'multiplier effect' on economy
    178–81
  post-project performance analyses
    206–7
  power to Delhi 139, 145
  regulation of waters of 68–70
  Report 149
  renovation, modernization and
    uprating of 143–4
  salient features of 72–6

structure of 72–3
transmission system of 146–7
Bhakra-Nangal Study Report 7
Bhakra spillway 139, 152, 154
Bhatia, Ramesh 9, 180, 181
Bilaspur, displaced people from 194
Bist-Doab canal, Punjab 57, 81, 82, 198
Black, Eugene 2
'black start power' 145
Borlaug, Norman 38
Boulder dam, USA 184
Boundary Commission, for Punjab,
    Bengal 21–3
    Award 38
British rule, in India, developments
    during 14–16

canal irrigation 14, 48–9, 171
Central Electricity Commission 135
Central Inland Fisheries Research
    Institute 51, 175
Central Rice Research Institute 101
Central Water Commission 92, 185, 186
Centre for Policy Research, study 6–8
Chandigarh city 176, 177
Chenab river 11–12, 30, 31, 36
Chopra, Kanchan 7
Chopra, M.R. 61
climate 42
Committee of Technical Experts,
    Report of 159n, 160
communication system 51, 53, 148
Constitution of India, on welfare of
    women 116
cropping pattern 88, 201–2

dairy development 203
dams, large 1, 4
    critics of 4
    and food production 122–3
    role of 131
    see also anti-dam lobby
Dane, Louis 17
Dantwala, M.L. 114

Dehar power station 33
Delhi, electrification of villages in 175
    Sultanate period, agriculture
        during 14
    water demands of 171–2
Delhi Jal (Water) Board 172
Department of Women and Child
    Development 117
Dhawan, B.D. 102, 127
displaced people, dams and 62, 63–4,
    74, 193–6, 213–14
    compensation for 194–5
    Relief and Rehabilitation package
        for 214, 215
    see also oustees
downstream reaches, canalization and
    embankment of 151–3
    flood problems in 158–61
drinking water supply 203–4, 211–13
    demand for 173
    problem of 41
drought 39, 40
    management 150
    -prone area 40
dryland farming 128

economy, Bhakra-Nangal Projects'
    multiplier effect on 178–81
education 204
electricity, access to, and
    consumption of 137, 167,
    208–10
Elephant Butte and Boulder dams of
    the USA 184
employment, Bhakra-Nangal Project
    and 177–82, 204
environment, impact of project on 39
evaporation 44

famines 39–40, 210
    Bhakra-Nangal Project and
        immunity from 165–6
Famine Commission, 39
    Reports of 40

farm mechanization, and
  unemployment 115–16
female disadvantage, and Green
  Revolution 116–20
  see also women
Ferozpur 22–3
fisheries 51, 175
Five Year Plan, First 136, 192
  Second 137
  Third 137
  Seventh 113
  Tenth 114
  Document 111
flood(s) 49–50, 152–3, 159, 160–1
  forecasting and warning system 158
  management 9, 149–64
  moderation 152, 154, 156, 158, 162
  regulation 156–8
  upstream flash 155–6
'Food for Work' programme 167
food crops, yield of 112
food grains, area under 125–7
  import of 96–8
  production 109, 110, 111–13,
    120, 122–34
  productivity 127
  yield 127

Gandhi, Indira 101
Ganguwal powerhouse 144, 148
geology, of the dam region 47–8
Ghaggar canal 80, 84
Ghaggar river 199
Gill, R.S. 61
Giri, V.V. 140
Gobind Sagar lake 62, 176
Gobind Sagar reservoir 159
greenhouse gases, emission of 5
'Green Revolution' 9, 95, 98, 113, 114,
    120–1, 131, 166, 178, 196, 202
  and female disadvantage 116–20
  impact of 104, 109
  in Punjab and Haryana 98–104
  technology 115

Grey canal 80, 83
groundwater, canal water and 201
  power for pumping 200
Gulhati, N.D. 91, 212
Gwyther, F.E. 19

Handa, C.L. 61
Hansli canal 15
Haramosh mountains 11
Harappan civilization 13
Harike barrage 26
Harris-White, B. 119
Haryana, climate of 42
  economic development of 17
  green revolution in 98–102, 104
  humidity in 45, 46
  irrigation coverage/development in
    89, 199–200
  land use in 103, 132
  physiography of 41
  population of 173
  poverty ratio of 168
  rainfall in 42, 43
  rice production in 108–32
  rural electrification in 174
  salinity in 94
  soils of 47–8
  temperature in 44
  water logging in 90–1
  wheat production in 108, 111,
    131
  wind speed in 45
Hashim, S.R. 129
Haveli canal 16
health status, improved 203–4
High Level Committee on Floods
  (1957) 50
high yielding variety (HYV) seeds,
  of rice, wheat 98, 100, 101, 114,
    115, 120, 166, 196
Himachal Pradesh, climate of 42
  physiography of 41
  resettlement of displaced people
    from Bhakra dam sites in 63–4

rural electrification in 175
soils of 47
Himalayas 11, 41, 47
Hirakud reservoir 191
Hisar, resettlement of displaced
  people 77–8, 195, 214
Hoovar dam 1
hydroelectric power, from Bhakra-
  Nangal project 9, 135–48, 193
  installed capacity of 75
HYDAC system 191

Iliff, William 31
India, food problem in 38
  Partition of 20–2 38, 80
    and migration of people 25–6
  post-Partition 38–55
  water dispute with Pakistan 27–8
  wheat import from USA 96–7
Indian Independence Act, 1947 21, 25
Indus basin, development of 8, 11
  headworks on rivers of 16–17
  irrigation development in 13, 25
  till 1947 11–23
  after 1947 25–37
Indus Basin Development Fund, 36
  World Bank assistance 34
Indus Commission 30
Indus river 11, 30, 36
Indus system, rivers of 12–13, 25, 41
Indus Water Treaty 9, 28–31, 34,
  91, 100
  negotiations 212
industries, development of 173–4, 205
infrastructural developments 204–5
Institute of Economic Growth 127
Integrated Drought and Flood
  Management 210–11
integrated flood management 210–11
International Commission on
  Irrigation and Drainage (ICID) 3
International Commission on Large
  Dams (ICOLD) 3
International Energy Outlook 2004 138

International Engineering Company
  (IECO), USA 135
International Hydropower Association
  (IHA) 3, 8
International Rice Research (IRRI) 101
International Rivers Network (IRN) 3
inundation canal system 13, 14
irrigation/irrigated, from Bhakra-
  Nangal 2, 15, 75, 80–94
  agriculture 127, 129, 196, 197
    and increased productivity
      95–121
  area under 165
  beneficiaries, from Bhakra-Nangal
    project 196–200
  development of 13, 88, 89
Irrigation Commission 40, 41
  First 15, 166
  Second 166
  1972 Report of 90, 91

Jhelum river 11, 30, 31, 36
Johnson, B.M. 61

Karakoram ranges 11
Kashmir, accession to India 25
Katoch, S.C. 61
Khan, Liaqat Ali 27
Khan, Mohammed Ayub 31, 91
Khanna, R.L. 61
Khungar, S.D. 61
Kennedy 91
Khosla, A.N. 19, 23, 26, 61, 184
Kinnaur district, Himachal Pradesh,
  flash floods in 155
Kirthar ranges 11
Kishanganga 11
Kol dam project 191
Kotla powerhouse 144
Krishna Raja Sagar lake, Mysore 176
Kumar, B.N. 161

labour, Bhakra-Nangal project and
  migrant 181–2, 203

Land Acquisition Act 1894 63, 64
land use/utilization 48–51
  in Haryana 132
  in Punjab 132
Lilienthal, David E. 28
livestock, and milk production,
  Bhakra-Nangal project and 169–71
Lower Bari Doab canal 16, 166
Lower Chenab canal 15
Lower Jhelum canal 15
Ludhiana, industries in 173–4
  population of 173

Madhopur-Beas Link 34
Makhu canal 26
malaria, measures against 54–5, 66
Malik, R.P.S. 180
Manasarovar lake 11, 12
Mandela, Nelson 5
Mangala dam, Pakistan 34, 36
Mauryan rule, public irrigation works
  during 13
medical facilities and health status, of
  project region 54
milk production, in Punjab, Haryana
  and Rajasthan 169–71
Minhas, B.S. 7, 100
Ministry of Environment and
  Forests 39
Moon, Penderel 22
Mountbatten, Louis 21, 23
Mughal rule, irrigation and water
  supply works during 14

Nangal barrage 17, 57, 58, 59, 158,
  213, 216
Nangal hydel channel (NHC) 57
  powerhouses on 136, 138, 139
Nangal hydel powerhouse 137
Nangal project 57–8
  irrigation system of 57–8
Nangal river 151
  outflow from 156
Narmada dam 4

Nathpa-Jhakri Project (NJP) 155,
  189, 191
National Commission on Agriculture
  (NCA) 93
National Commission for Integrated
  Water Resources Development
  (NCIWRD) 129–30
National Flood Commission 163, 206
national flood policy statement 1954 50
National Policy on Resettlement and
  Rehabilitation 215
National Water Policy 41, 69, 173
Nehru, Jawaharlal 23, 27, 31, 56,
  57, 58, 60, 86, 87, 137, 140
Nicholson, H.W. 19, 51
Nickel, F.A. 20
Nippon Foundation, Japan 6, 8
Northern Electricity Grid 9
Northern Regional Grid of India 139,
  144–6

Oustees, allotment of land to Bhakra,
  in Hisar district 77–8

Pakistan 9, 11, 12
  creation of 21, 22
  note on Sutlej water 86
  post-Independence developments
    in 34–6
  water dispute with India 27–8
Palta, B.R. 51, 61
Panjnad headworks 16
Para canal 15
PEPSU 32, 38
physiography 41–2
pisciculture 175
Planning Commission 136, 137, 166,
  171, 192, 207, 208
  'Working Group on Flood
    Control' 163
plant species 50
Pong dam 34
post and telegraph 53–4
poverty, alleviation, Bhakra-Nangal

Project and 166–7
    ratio 168
power, generation 72
    planning for 207–10
    demand for 144–5
    production, load estimate and
        136–8
power line carrier communication
    (PLCC) 148
Project Reports of 1919, 1939–42,
    1951 135
public distribution system (PDS),
    poverty alleviation and 167
Public Law (PL) 480, USA 96
Punjab, climate of 42
    crop yield in 49
    cultivable area in 48
    division of 38, 80
    economic development of 179
    floods in 49
    green revolution in 98–102,
        104, 115
    humidity in 45, 46
    irrigation coverage/development
        in 48–9, 89, 197–9
    land use in 103, 132
Punjab
    partition of 21, 22
    physiography of 41
    population of 173
    poverty ratio of 168
    power shortage in 20
    rainfall in 42, 43
    reorganization of 32
    rice production/yield in 106,
        132
    rural electrification in 174
    salinity in 94
    soils of 47
    temperature in 44
    water logging in 90
    wheat production/yield in 105,
        111, 131, 134
    windspeed in 45

Punjab Boundary Commission 21
    Award 38
Punjab Electricity Department
    135
Punjab Irrigation Department 59
Punjab Partition (Apportionment
    of Assets and Liabilities)
    order 1947 27
Punjab Partition Committee 27
Punjab Reorganization Act 1966
    32, 68
Punjab State Electricity Board
    (PSEB) 146

Radcliffe, Cyril 21, 38
Radcliffe Award 22, 23, 25, 38
railways 53
rainfall 40, 42
    and floods 160
rain-fed irrigation 199
Raj, K.N. 7
Rajan, Irudaya 117
Rajasthan, climate of 42
    drinking water problem in 41
    green revolution in 99
    irrigation coverage/development
        in 89, 200
    physiography of 42
    rainfall in 42
    rainfall, temperature and
        windspeed in Gangasagar
        area 45, 46
    rural electrification in 175
Ram, Chhotu 19
Randhawa, M.S. 98, 99
Rangachari Committee 163
Rao, C.H. Hanumantha 113
Ravi river 11, 12, 15, 30, 35
Ravi-Beas-Sutlej system 171
    integrated operation of Sutlej
        and 67–8
Ravi-Beas Water Disputes Tribunal
    1986 32
Razavi, Shahra 119

regional imbalance 113–14
research institutes/universities 202–3
reservoir schemes, for flood control
    149
rice, procurement 13
    production 101, 102, 104, 120–1
    production/yield, in Haryana
        108, 132
    in Punjab 106, 132
roads 53
Rockefeller Foundation 101
Roopnagar headworks 81, 82
Roy, Arundhati 4
rural electrification, Bhakra-Nangal
    project, and 174–5
RVP Sutlej scheme 190

SAARC 116
Sailab (irrigation type) 13
Sain, Kanwar 61
Saith, R. 19
salinity, water logging and 88, 90–4
Sanjay Vidyut Pariyojana, on Bhakra
    189
Sardar Sarovar project 211
    controversy over 2
Sarsuti canal 80, 84
Savage, J.L. 20
sedimentation, in Bhakra reservoir
    183–91
    effects of 185
    mitigation measures for 189–91
Seervai, H.M. 23
Sengupta, Nirmal 123–5, 128
sex ratio, in India 116, 117
Shastri, Lal Bahadur 97
Shiva, Vandana 118, 119
Sidhnai canal 15
siltation, problem of 186, 189, 190–1
Singh, Himmat 94
Singh, Jagman 53
Sirhand canal system, Punjab 15, 26,
    67, 80–4, 131, 165, 197
Sirsa river 160

Siwaliks 12, 41, 47
Slocum, M.H. 61
Soan *nadi* 151
social and economic conditions 203–5
social inter-class disparities 114–15
Sohag canal 15
soils 47–8
Soviet Union, assistance from 143
Srinivasan, T.N. 100
Sriramsagar reservoir 191
standard of living, Bhakra-Nangal
    project and improvement in 205
State Waterlogging Board and
    Waterlogging Conference 1928 90
Subramaniam, C. 97, 98, 100
Sudha, S. 117
Sukkur barrage 16
Suleimanki ranges 11
Sutlej basin 41
Sutlej and Ravi-Beas system 67–8
Sutlej river 11, 12, 15, 16, 22, 26,
    30, 33, 34, 36, 56, 57, 58, 60,
    81, 86, 138, 139, 149, 151, 164,
    167, 198
    canalization of 152
    channel 151
    embankment 155
    flood 160
    silt load in 9
Sutlej Valley Project 16, 19
Swan river 160
Swaraj, Sushma 117

Tarabela dam, Pakistan 36
Tehri dam project, controversy over 2
temperature 44
Thal canal 16
Thein (Ranjit Sagar) dam 32
tourism, Bhakra-Nangal project and
    176–7
town and country planning, Bhakra-
    Nangal project and 177
transpiration 44
'Triple canals', West Punjab 16

tube-wells 178, 198
    irrigation 120
Tungabhadra reservoirs 191
TWCWM 6, 8

Ukai reservoir 191
unemployment, farm mechanization
    and 115–16
United Nations Women's Decade
    116
Uppal, H.L. 90
Upper Bari Doab canal (UBDC)
    system 15, 27, 131
Upper Jhelum canal 16
Uttar Pradesh, green revolution
    in 99

Vaidyanathan, A. 124, 128

Warabani system 203
wasteland, Bhakra-Nangal dam and
    reclamation of 167, 169
water, conservation technologies 202
    dispute between India and
        Pakistan 25, 27–8
    quality of, from Bhakra dam
        50–2
    resources development schemes
        213, 216
    supply from Bhakra-Nangal 171–3
    users' association 203

water logging, and salinity/alkalinity
    88, 90–4
watershed management 189, 190
Wavell, Lord 21
well irrigation 16, 48
Western Jamuna canal 83, 131,
    165, 172
wheat, procurement 132–4
    production/yield in Punjab and
        Haryana 102, 104–5, 107,
        111, 120–1, 131
wind speed 45
women, Bhakra-Nangal project and
    advantage for 178
    green revolution and
        discrimination against 118–19
    welfare of 116
World Bank 3, 29, 30, 31, 180, 208
    on Indus solution 67
    Performance Review 123
World Commission on Dams
    (WCD) 2–6, 122, 123
    Report 5, 6, 20–3, 179
Working Group on Waterlogging
    1992, Report of 91–3
Wular lake 11

Yamuna river 14, 41, 171, 172
Year of the Girl Child 116

Ziegler, Philip 23